**Introduction to Statistical
Process Control**

Introduction to Statistical Process Control

Muhammad Aslam

Aamir Saghir

Liaquat Ahmad

Registered Office
John Wiley & Sons, Inc., 111 River Street, Hoboken, NJ 07030, USA

Editorial Office
111 River Street, Hoboken, NJ 07030, USA

For details of our global editorial offices, customer services, and more information about Wiley products visit us at www.wiley.com.

Wiley also publishes its books in a variety of electronic formats and by print-on-demand. Some content that appears in standard print versions of this book may not be available in other formats.

Library of Congress Cataloging-in-Publication Data is applied for

9781119528456

Cover Design: Wiley
Cover Image: © MF3d/Getty Images

Set in 9.5/12.5pt STIXTwoText by SPi Global, Pondicherry, India

10 9 8 7 6 5 4 3 2 1

To my parents and my wife, Nagina Aslam   – M.A.

To the memory of my father   – A.S.

To my parents, without whom none of my success would be possible   – L.A.

Contents

About the Authors

Muhammad Aslam is a full professor of statistics in the Department of Statistics, King Abdulaziz University, Jeddah, Saudi Arabia. He did his MSc in Statistics (2004) from GC University Lahore with Chief Minister of the Punjab merit scholarship, MPhil in Statistics (2006) from GC University Lahore with the Governor of the Punjab merit scholarship, and PhD in Statistics (2010) from National College of Business Administration & Economics (NCBAE), Lahore, under the kind supervision of Professor Dr. Munir Ahmad. He also worked as research assistant in the Department of Statistics, GC University Lahore from 2006 to 2008. Then he joined the Forman Christian College University as a lecturer in August 2009. He worked as an assistant professor in the same university from June 2010 to April 2012. He worked in the same department as an associate professor from June 2012 to October 2014. He worked as an associate professor of statistics in the Department of Statistics, Faculty of Science, King Abdulaziz University, Jeddah, Saudi Arabia, from October 2014 to March 2017. He taught summer course as a visiting faculty of statistics at Beijing Jiaotong University, China, in 2016. He has published more than 390 research papers in national and international well-reputed journals. He is the author of two books. He received Meritorious Service Award in research from NCBAE, Lahore, in 2011. He also received Research Productivity Award for the year 2012 from Pakistan Council for Science and Technology. He was the second-listed statistician in the Directory of Productivity Scientists of Pakistan 2013. He was the first-listed statistician in the Directory of Productivity Scientists of Pakistan 2014. He got 371th position in the list of top 2210 profiles of Scientist of Saudi Institutions 2016. He received King Abdulaziz University Excellence Awards in scientific research.

Neutrosophic statistical quality control (NSQC) was introduced for the first time by Professor Muhammad Aslam. He is also the founder of neutrosophic inferential statistics. He developed neutrosophic statistics theory for the inspection and process control. He originally developed theory in these areas under the neutrosophic statistics. He extended the classical statistics theory to neutrosophic statistics originally in 2018.

Aamir Saghir is currently working as an associate professor of statistics in Mirpur University of Science and Technology, Azad Jammu and Kashmir, Pakistan. He obtained his PhD degree in probability theory and mathematical statistics from Zhejiang University, China, in 2014. He obtained his MS degree in statistics from Quaid-i-Azam University, Islamabad, Pakistan, in 2008. His current research interests include statistical process monitoring and probability modeling. He published more than 34 articles in internationally well-reputed International Scientific Indexing journals. He also served as a peer reviewer for many international journals in the field of statistics and mathematics.

Liaquat Ahmad is currently working as an associate professor in the Department of Statistics and Computer Science, University of Veterinary and Animal Sciences (UVAS), Lahore, Pakistan. He has published 35 research articles in highly reputed international impact factor journals. He has also reviewed more than 30 articles of international journals in the field of statistics. His areas of interest are quality control charts, acceptance sampling plans, experimental design, biostatistical data analysis, and statistical softwares.

Preface

This book is about the use of modern statistical methods for quality control. It provides comprehensive coverage of the subject from basic principles to state-of-the-art concepts and applications. The objective is to give the reader a sound understanding of the principles and the basis for applying them in a variety of situations. Although statistical techniques are emphasized throughout, the book has a strong engineering and management orientation. Extensive knowledge of statistics is not a prerequisite for using this book. Readers whose background includes a basic course in statistical methods will find much of the material in this book easily accessible.

The book originally grew out of our teaching, research, and consulting in the application of statistical methods for various fields, particularly for industries. It is designed as a textbook for students enrolled in colleges and universities, who are studying engineering, statistics, management, and related fields and are taking a first course in statistical quality control. The basic quality control course is often taught at the junior or senior level. All the standard topics for this course are covered in detail. We have also used the text materials extensively in programs for professional practitioners, including quality and reliability engineers, manufacturing and development engineers, product designers, managers, procurement specialists, marketing personnel, technicians and laboratory analysts, inspectors, and operators.

The book contains eight chapters. Chapter 1 is an introduction of history and background of control charts. This chapter also includes descriptive statistics, the basic notions of probability and probability distributions, and types of control charts. These topics are usually covered in a basic course in statistical methods; however, their presentation in this text is from the quality engineering viewpoint. It describes that quality has become a major business strategy and that organizations that successfully improve quality can increase their productivity, enhance their market penetration, and achieve greater profitability and a strong competitive advantage.

Chapter 2 presents the Shewhart-type attribute control charts for counts. It highlights the importance of attribute data and presents the proper use of statistical methods to implement quality control. Concept of dispersed data is introduced, and control charts used for monitoring over- or under-dispersed data are discussed along with application in industries.

In Chapter 3, control charts for monitoring the variable data processing are presented with basic concepts and implantation on real data sets. These Shewhart control charts certainly are not new, but its use in modern-day business and industries is of tremendous value. Chapter 4 contains the new idea of control charts for multiple dependent state (MDS) sampling. The MDS sampling showed the efficiency of the attribute control chart over the traditional Shewhart attribute control chart in terms of average run length. The use of MDS sampling in the area of control chart has increased the sensitivity of the control charts to detect a small shift in the manufacturing process. For decision-making, it uses the current subgroup information and previous subgroup information to make the decision about the state of the process. In Chapter 5, exponentially weighted moving average (EWMA) control charts using repetitive group sampling scheme are introduced. The methods for EWMA-based control charts for a variety of situations, such as the average and the dispersion monitoring charts, single sampling, double sampling, multiple sampling, sequential sampling, repetitive sampling, ranked set sampling, and the MDS sampling charts have been developed in this chapter.

Chapter 6 presents the different sampling schemes used to construct the control charts. Some of these sampling schemes are very simple to develop and understand, while some schemes are much complex to develop and understand. Each of the sampling schemes has advantages and disadvantages; therefore the quality control personnel can select according to the situation and the available resources. The use of modern statistical software has made a very simple task of developing a control chart for the non-statisticians' quality control personnel as there is no need to develop and understand the complex sampling schemes. In Chapter 7, memory-type control charts for monitoring attributes, such as the cumulative sum chart, the EWMA chart, and the moving average charts, are given with application in industries. Chapter 8 contains the material related to multivariate process control schemes. Nowadays, in industry, there are many situations in which the simultaneous monitoring or control of two or more related quality process characteristics is necessary. Monitoring these quality characteristics independently can be very misleading.

Throughout the book, guidelines are given for selecting the proper type of statistical technique to use in a wide variety of situations. Additionally, extensive references to journal articles and other technical literature should assist the reader in applying the methods described.

Muhammad Aslam
Aamir Saghir
Liaquat Ahmad
May 2020

Acknowledgments

The writing of this book was a challenging task and needed many months of concerted efforts, which involved long working hours for which our families sacrificed tremendously over this long period of time. We thank them for their patience and understanding.

We first recognize Elisha Benjamin, the project editor; Kathleen Santoloci, associate editor; and Mindy Okura-Marszyck, senior editor, at Wiley, for providing many invaluable advice and help during the writing of this book. They have always been very kind and prompt with their replies for our queries. We like to sincerely thank both of them from the bottom of our hearts.

Professor Muhammad Aslam would like to thank Professor Munir Ahmad (Late) of NCBAE, Lahore, Pakistan, who is his a professor, mentor, coauthor, and a friend who had introduced him to this line of research. He would also like to thank Professor Chi-Hyuck Jun of Pohang University of Science and Technology, South Korea, who is his coauthor, mentor, and friend. He thank the Department of Statistics, Faculty of Science, King Abdulaziz University for providing excellent research facilities and his colleagues in the university Dr. Saeed A. Dobbah, Professor Ali Hussein AL-Marshadi, Professor Mohammed Albassam, Professor Kushnoor Khan, and his colleagues from UVAS, Lahore, Pakistan, Professor Muhammad Azam, Professor Liaquat Ahmad, and Dr. Nasrullah Khan for their constant encouragement.

The authors gratefully thank all the publishers from which some tables and figures were reproduced in the book.

Finally, the authors are grateful to their parents and their families and their respective wives.

1

Introduction and Genesis

1.1 Introduction

Quality improvement is a continuous process adopted in all business activities for two purposes: to compete the market and to maximize the profits. It is a competitive tool used for improving and controlling many organizations, transportation, health care, and government agencies. A goods and services providing agency delighted its customers by improving and controlling its quality, which dominates over its competitors. Shewhart control charts are used for this purpose, but according to experts of quality control, the process should not be disturbed until sound statistical inference evidence is used for indicating that the process is misbehaving. Without statistical evidence the process should not be modified. Consistent quality improvement can be sustained not only by modification of the process but also by redesigning of the process. The process design can only be changed after the full satisfaction of the quality control personnel who is running the process fully in control state (Figure 1.1).

Control charts are constructed using a reasonable size normally of 5 to 25 units of rational subgroups, periodically, from the running process. The statistically calculated values of these subgroups are posted on the limits of thus calculated values from that process. The posting of these subgroups indicates the fluctuations caused by the common or/and special causes of variations of the process under study. When all these subgroups are commingled with, then these values do not provide us the required information from the process as most of the information will be lost.

This book is written with the objective to help the quality control personnel to use statistical tools and techniques for monitoring and improving the quality of the product. Different techniques are available for different situations. The proper choice of the available techniques is the required competency of the quality control personnel.

Preamble: Some basic concepts regarding quality control and statistics are described before understanding the concepts and techniques in the forthcoming chapters.

Introduction to Statistical Process Control, First Edition. Muhammad Aslam, Aamir Saghir, and Liaquat Ahmad.
© 2021 John Wiley & Sons, Inc. Published 2021 by John Wiley & Sons, Inc.

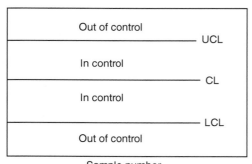

Figure 1.1 A typical control chart.

Quality improvement is an ever-present marvel. Originally developed techniques for the manufacturing environment are applied for the ever-increasing competition of the markets. These techniques not only are meant for the manufacturing processes but also have a wide scope and range in different areas from health care to education to government services. Statistical process control (SPC) is collection of techniques and methods for thinking about the data. The apparent utilization of these techniques may be to monitor the diameter of the bolt being produced on a manufacturing unit in bulk. Therefore, the collection of the sample from the assembly line to declaring the process whether it is in control or out of control is literally performed through SPC. Adopting these techniques will indicate the manufacturing deterioration of the bolt production may be caused by raw material or the fault in the production steps.

The SPC notion instigated during the twentieth century when Walter A. Shewhart float the idea of control chart during 1924. Another important technique of SPC is the acceptance sampling plans, which were introduced by Dr. H. F. Dodge and H. G. Roming in 1928 at Bell Laboratories. The variations in the products are important to designate when they deviate from the affordable/acceptable levels of variations. These variations are based on the principle of random process and monitored through control charts. The common steps used for the construction of a typical control charts can be listed as:

1) Decide a variable continuous or discrete to be monitored for the product. We consider a continuous variable here.
2) Calculate the mean of the targeted variable and the grand mean being used as central line (CL).
3) Calculate the standard deviation of the targeted variable.

4) Calculate the upper control limit (UCL) and the lower control limit (LCL) of the targeted variable with the deviation of three-sigma from the grand mean.
5) Plot the means collected from the targeted variable on the chart with the mean − 3sd and mean + 3sd sigma limits defined in Step 4.
6) On plotting the means there may be some points falling outside the UCL or LCL. If this exists, then probe the matter for the reasons behind these out-of-control values.
7) Revise the control chart for the new CL, UCL, and LCL after discarding the disturbing means.
8) Plot the means of the next collected data on the constructed limits and decide to declare the process as within control or out of control.

The SPC techniques are commonly used in the health care sector. For example, different blood parameters of the collected samples can be monitored for possible unusual changes occurred in the targeted variables. The quick detection technique of exponentially weighted moving average (EWMA) has been frequently used for efficient monitoring of the data consisting of various parameters of blood test in cats and dogs. It has been shown that the application of EWMA technique on the blood parameters is an effective method for such data.

The SPC techniques are also very commonly used in the human resource management sector. For example, the monitoring of misconduct of the workers, unrest among the employees of an organization, racial discrimination in a community, women harassment incidence in an office are being monitored by the application of the SPC techniques. Such parameter allows the managers for observing the prompt changes in the current status quo and any deviation of the prevailing policies of the organization.

The important concern of the SPC is based on the use of sampling techniques. The already developed methodologies are the single, double, with plenty of literature available in the books of SPC, but the repetitive sampling and multiple dependent state sampling schemes, which are the most commonly used in quite near past, are explained in this book. Many other sampling schemes, which are also tenderly accepted in SPC, are the probability to ratio sampling scheme, the ranked set sampling, and fast initial response set sampling scheme.

1.2 History and Background of Control Charts

Most of the SPC techniques being used nowadays are the techniques developed during the twentieth century. The control chart technique was basically introduced by Walter A. Shewhart in 1920s during his services rendered to Bell Laboratories. Two types of variations (common and special causes of variations) in the products of the manufacturing units were pointed out in an internal memo of about one page length written by Walter A. Shewhart on 16 May 1924. That one page consisted of text, and one-third of the page was utilized for the diagram of the shape showing UCL and LCL, which we are using nowadays as control chart. Two renowned quality control experts, connoisseurs, aficionadas, buffs, and polishers of the statistical methods applied to the quality management are Shewhart and Deming who wrote a book, *Statistical Methods from the Viewpoint of Quality Control*, in 1939, which is helpful and supportive even today as it was then (Oakland, 2008, p. 14). After the defeat of Japan in Second World War, Dr. William Edwards Deming (Shewhart's boss), basically an American statistician, engineer, professor, author, and management consultant, helped Japanese companies to improve the quality of the products by focusing on the monitoring and diagnosing of variability. The causes of variation were focused particularly in the manufacturing industry to improve performance through quality management system and SPC. Many companies in the world adopted the Deming philosophy and attained success swiftly in the years to come. The popularity of SPC for the success of manufacturing industry delivered the message to suppliers of raw material, goods and services providing organizations, and all the companies related to the manufacturing industry about the enormous potentials of the SPC in terms of market share, maximizing profits, and reducing rework of products. As a result a colossal demand for quality control techniques, SPC experts, and utilization of computer technology was created to survive in the world.

In 1935 the British Standards Institution introduced modification as the three-sigma control limits were replaced with the limits based on the percentiles of normal distribution.

Since the early 1980s, the US industry improved the quality substantially due to theories developed by the researchers. Dr. Genichi Taguchi, Dr. Joseph M. Juran, Dr. Deming, and Philip Crosby assisted the manufacturing industry magnificently for improving the product quality. A new boast was injected to the industry by Dr. Genichi Taguchi by introducing new concepts in experimental design, robust design, and loss function. The quality system of ISO 9000 and QS 9000 was introduced by the United States in the early 1990s.

Most of the emphasis on the execution of the control chart has been seen in the second half of the twentieth century when the monitoring of small shifts in the production process was focused by announcing two methodologies. These methodologies were the cumulative sum introduced by E. S. Page in 1954 and the EWMA by S. W. Roberts in 1965. Another direction was introduced by R. E. Sherman in 1965 when he hovered the idea of repetitive group sampling. The technique of the repetitive sampling was readdressed by S. Balamurali and C. H. Jun in 2006. Currently, a plenty of literature can be seen on the application of repetitive group sampling technique with the major contributions of M. Aslam and L. Ahmad (Ahmad, Aslam, & Jun, 2014; Aslam, Srinivasa Rao, Ahmad, & Jun, 2017).

1.3 What Is Quality and Quality Improvement?

Quality may be defined as the characteristic of a good or service that can fulfill our desire. In other sense, it is defined as the fitness to use. This quality of product is used as a powerful tool for deciding/selecting from many competing products. This quality plays a vital role in the growth, business success, and to face the competing market. It is a widespread phenomenon as the quality may be an individual, a departmental store, a food chain, a cellular company, a transport company, etc. The quality may be classified as the quality of the design and the quality of the conformance. The quality of design may be described as the goods and the services produced in various designs and levels. The quality of the product like other corporate matters must be reviewed continually in the light of current settings. Two of the most famous authors, Shewhart and Deming, worked for the quality and management of the quality using statistical theories and concepts.

The quality of conformance depends on several factors, such as specifications of the design, selection of manufacturing process, types of process control, etc. Garvin (1987) described eight dimensions of quality. These dimensions are explained as follows:

i) Performance

How much the product is worth for the intended job? The product or services are evaluated for its specific functions. If the desired functions are performed well, then the performance of the product is high.

ii) Reliability

How often does the product fail to perform? When the use of the product is fit for a long time, then it is said the product is reliable for a long time. When a product requires repair or change frequently, then it is said to be unreliable. For example, if the use of plug-in motor car engine lasts for long time, then its reliability is high; otherwise it is said to be unreliable.

iii) Durability

How long the product serves? This is the most common property desired by the customers for any product. The customers want to purchase the products which serve them for longer lifetimes. Particularly, this dimension is applicable in automobile, electronics, communication products, etc.

iv) Serviceability

How informal is it to get repair? Wear and tear is common in any product. The quality of any product is fit to the consumer if it is easy to get its service. The purchasing decision of the product is directly related to the serviceability of that product. For example, the mobile manufacturing companies normally provide after-sales service of the mobile. Now if any problem is met to the user, then most of the companies provide the replacement of the mobile and some companies pronounce time for its repair as they send the mobile in the workshop, which may take weeks or even months to deliver it back to the customer.

v) Aesthetic

How beautiful the product looks? The look of the product plays an important role in determining the quality of the product. There are many factors that are directly related to the aesthetic of the product, for instance, the color, design, shape, model, style, packing, and such other sensual features. For example, in automobile industry the color and design of a vehicle play a significant role toward its quality.

vi) Features

What extra functions the product can do? The customers have the tendency toward the articles having some extra features which others brands do not have. These added features are provided in addition to the original or main functions for which that article has been given. For example, if a mobile phone manufacturing company introduces a function of converting voice to text form while others do not have this function, then this is quality added function.

vii) Perceived Quality

What is the repute of the product? Usually, customers are associated with those companies or products with which their past experience is pleasing and satisfying. Most of the customers attach the experiences of such other products while deciding to purchase or hiring a new one product or service. Normally the brand once selected urges the customers to decide for the next time. For example, consider an electronic company manufacturing a wide range of electronic goods like air conditioner, deep freezer, cooking range, washing machine, television sets, LEDs, mobiles, etc. If a customer uses a mobile phone of that company and he feels zero problem and fully satisfied with its working and after-sales services or experience of warranty claim, then absolutely he will prefer to purchase such other items like air conditioner, deep freezer, cooking range, washing machine, television sets, and LEDs of that company.

viii) Conformance to Standard

How much the product is made according to the set standard? The product quality is very much related to correctness of the designed item with the final product. Any product is designed first based on so many parameters and then sent to the manufacturing unit for production. Any difference between the design and the final shape of the product affects the quality of the product.

In connection with these concepts, the quality may be defined by either the characteristic of fitness for intended purpose or the conformance to specification. To achieve target explained in the chief definition, which is much broader concept, is not so easy task, but a continuous monitoring of manufacturing and management system and the later concept which is related to manufacturing phase is monitored using the SPC technique.

The long-term effect of the quality is a desirable target for the administration of the business. Only obtaining the quality or service of the product is not enough, but keeping it up is a continuous process, which demands for financial allocation. Now we analyze the cost aspects for maintaining the quality of the product. There are three aspects of cost burden associated with the quality of the product. These cost aspects are the real efforts toward attaining the quality of the product. It is sole responsibility of the management to keep the balance between the quality and the cost incurred for maintaining such quality. A competent analysis of the costs of quality will lead to the objective of the administration.

The quality costs are the same costs as happened other costs like production cost, distribution cost, promotion cost, maintenance cost, sales cost, design cost, etc., which is a significant management tool for monitoring and assessing overall effectiveness of the treated quality.

Types of Quality-Related Costs

The quality-related costs can be divided into three types:

a) Prevention Cost

These are costs incurred before the production or operation. These are associated with the design, implementation, and the maintenance of the quality management system costs (Garvin, 1987):

Design Costs: The setting and determination of the required design and corresponding specifications, the incoming material, etc.
Quality Planning Cost: The cost incurred on the formation of quality, process control, inspection, and supervision.

Quality Assurance Cost: The establishment of a system for the overall management of the product or service to ensure whether the desired quality is maintained.

Cost on Inspection Equipment: The equipment used for the inspection of the design and corresponding specifications, the incoming material, etc.

Training Cost: The training of the supervisors, operators, managers, workers, etc. for the production of set quality and maintaining it for the future production.

Miscellaneous Cost: It includes the cost of general office management for the said quality, i.e. travel, communication, shipping, and clerical.

b) Appraisal Cost

These are the costs associated with checking whether the products are going to be produced according to the specifications. Normally, these costs are associated with the evaluation of goods and services administered by the producers or the purchasers as testing of raw materials, in-between process checking, etc. These costs include the following (Garvin, 1987):

Verification Costs: The cost associated with the verification of raw materials, verification of process settings, and final product verification.

Quality Audit Costs: The cost associated with checking of the management for its satisfactory working.

Inspection Kit Costs: The cost on the tools used for the inspection of the whole production process associated with the quality.

Vendor Scoring Cost: Different suppliers are associated with the production process. The cost incurring on the rating of these suppliers is included in the vendor scoring costs.

c) Internal and External Failure Cost

These are the costs associated with the products during its manufacturing that do not meet the designed standards; these failures are detected before transforming/delivering the products to the consumers. The failures/defective items till these are delivered to the consumers are known as the internal failures, while the after-sale failures/defective items are known as external failures (Oakland, 2008):

i) Internal Failure Costs

Scrape Cost: It is cost of items which do not meet to the specifications of quality. These items are declared as scrape when these are unable to rework as well as unable to use.

Reinspection Cost: When the items are reinspected after setting of all standards.

Rework Cost: The cost on the minor defects, which can be addressed only some small activity.

ii) External Failure Costs

Warranty Claims: The cost associated with the claims of defective items if these are delivered to the consumers.

Complaints: The cost associated with the complaints of the consumers when the products are not working properly.

Loss of Repute and Liability: Due to the sale of defective items to the consumers, the administration and firm have to face the loss of repute and its whole liability, which brings huge financial burden for the firm.

Return of Products: Most of the times the customers are so fed up with the articles that they do not want to use it anymore and prefer to return.

1.4 Basic Concepts

1.4.1 Descriptive Statistics

Statistics is the science used for the collection, presentation, analysis, and the interpretation of data. Descriptive statistics aims to describe various aspects of the data obtained in any data collection activity. The descriptive statistics consists of the listing, summary statistics, and the graphing of the collected data. The important measures of descriptive statistics are the mean, the median, the mode, the geometric mean, the harmonic mean, range, quartile deviation, mean deviation, variance, and standard deviation.

i) The Mean

The mean is defined as the average of the data set. It can be defined as the value obtained by dividing the sum of the values by their numbers. It is denoted by $\overline{X}$. The formula for the calculation of the mean is

$$\overline{X} = \Sigma X/n$$

Example 1.1 The weights of 10 items selected from the assembly line are given as

24.3, 24.9, 23.6, 25.1, 26.7, 22.0, 21.8, 23.4, 24.1, and 20.8. Calculate mean of the weights:

$$\text{Mean} = \overline{X} = \Sigma X/n$$
$$= 236/10$$
$$= 23.60$$

The mean for the group data can be calculated as

$$\text{Mean} = \overline{X} = \Sigma fX \big/ \Sigma f$$

ii) The Median

The median is defined as the value that divides the data into two equal parts. It is denoted by $\widetilde{X}$. The median calculation for the ungrouped data can be made as

$$\text{Median} = \widetilde{X} = \left(n + \tfrac{1}{2}\right) \text{th value} \qquad \text{(for odd number of values)}$$

$$\widetilde{X} = \left[\left(^{n}/_{2}\right)\text{th} + \left(\left(^{n}/_{2}\right) + 1\right)\text{th value}\right]\big/_{2} \qquad \text{(for even number of values)}$$

Example 1.2 The weights of nine items selected from the assembly line are given as

24.3, 24.9, 23.6, 25.1, 26.7, 22.0, 21.8, 23.4, and 20.8. Calculate median of the weights.

Arranging the values in ascending order

20.8, 21.8, 22.0, 23.4, 23.6, 24.3, 24.9, 25.1, and 26.7.

$$\text{Median} = \widetilde{X} = \left(n + \tfrac{1}{2}\right) \text{th value} \qquad \text{as the number of values is odd}$$

$$= \widetilde{X} = \left(9 + \tfrac{1}{2}\right) \text{th value}$$

$$= \widetilde{X} = \left(9 + \tfrac{1}{2}\right) \text{th value}$$

$$= \widetilde{X} = (5)\text{th value}$$

$$= \widetilde{X} = 23.6$$

In case of even number of observations, the median is defined as the mean of the two most middle values.

Example 1.3 The weights of 10 items selected from the assembly line are given as

24.3, 24.9, 23.6, 25.1, 26.7, 22.0, 21.8, 23.4, 24.1, and 20.8. Calculate median of the weights.

Arranging the values in ascending order

20.8, 21.8, 22.0, 23.4, 23.6, 24.1, 24.3, 24.9, 25.1, and 26.7.

$$\widetilde{X} = \left[\left(^{n}/_{2}\right)\text{th} + \left(\left(^{n}/_{2}\right) + 1\right)\text{th value}\right]\big/_{2}$$

$$\widetilde{X} = \left[\left(^{10}/_{2}\right)\text{th} + \left(\left(^{10}/_{2}\right) + 1\right)\text{th value}\right]\big/_{2}$$

$$\tilde{X} = [(5)\text{th} + (6)\text{th value}]\big/_2$$

$$\tilde{X} = [23.6 + 24.1]\big/_2$$

$$\tilde{X} = [23.6 + 24.1]\big/_2$$

$$\tilde{X} = 23.85$$

The median for grouped may be defined as

$$\tilde{X} = l + \frac{h}{f}\left(\frac{n}{2} - c\right)$$

where l is the lower class boundary of the median class and c is the cumulative frequency.

The family of median consists of partitioning of observations into different parts. As median is defined as the value that divides the data into 2 equal parts, the observations may be divided into 4, 10, or 100 equal parts. These measures are known as the quantiles. When data are divided into four parts, then it is called quartiles. Upper and lower quartiles or Q_1 and Q_3 may be defined as

$$Q_1 = \left((n+1)\big/_4\right)\text{th value}$$

and

$$Q_3 = \left(3(n+1)\big/_4\right)\text{th value}$$

The decile is defined as the value that divides the data into 10 equal parts.

$$D_1 = \left((n+1)\big/_{10}\right)\text{th value}$$

The other values of deciles can be calculated accordingly as the subscript is multiplied with the numerator. For example, the sixth decile may be calculated as

$$D_6 = \left(6(n+1)\big/_{10}\right)\text{th value}$$

The percentiles may be defined as the value that divides the data into 100 equal parts:

$$P_1 = \left((n+1)\big/_{100}\right)\text{th value}$$

The other values of percentiles can be calculated accordingly as the subscript is multiplied with the numerator. For example, the 70th percentile may be calculated as

$$P_{70} = \left(70(n+1)\big/_{100}\right)\text{th value}$$

The formula for the grouped data can be constructed accordingly.

iii) The Mode

The mode is defined as the most repeated value in any data set. It is denoted by $\hat{X}$. If there is no repetition in the data, then there will be no value of this measure. This is the single measurement whose results may be no value of mode, one value of mode, two values of the mode, or more values of mode in any data set.

The mode for the grouped data may be calculated as

$$\hat{X} = l + \frac{f_m - f_1}{(f_m - f_1) + (f_m - f_2)}(h)$$

iv) The Range

The range is defined as the difference between the maximum value and the minimum value of the data set:

$$\text{Range} = X_{max} - X_{min}$$

or

$$\text{Range} = X_n - X_0$$

The range for the grouped data may be calculated as

$$\text{Range} = \text{Upper class boundary of the highest class} -$$
$$\text{Lower class boundary of the first class}$$

v) The Quartile Deviation

The quartile deviation is defined as the half of the difference between the third and the first quartiles of the data set. The formula for the quartile deviation is defined as

$$QD = (Q_3 - Q_1)/2$$

It is to be noted that all the measures of dispersion are positive.

The coefficient of QD can be calculated as

$$\text{Coefficient of QD} = (Q_3 - Q_1)/(Q_3 + Q_1)$$

vi) The Mean Deviation

The mean deviation is defined as the average of the deviations from the mean or median; the deviations are taken without algebraic sign. So the average deviation calculated from mean is known as the mean deviation from mean and is defined as

$$MD = \Sigma|X - \overline{X}|/n \text{ is the mean deviation from the mean}$$

And the mean deviation from median and is defined as

$$MD = \Sigma\left|X - \widetilde{X}\right|\Big/_n$$ is the mean deviation from the median

And for the grouped data these measures can be defined as

$$MD = \Sigma f\left|X - \overline{X}\right|\Big/_{\Sigma f}$$ is the mean deviation from the mean

And the mean deviation from median and is defined as

$$MD = \Sigma f\left|X - \widetilde{X}\right|\Big/_{\Sigma f}$$ is the mean deviation from the median

The coefficient of MD is defined as

$$\text{Coefficient of mean deviation} = MD\Big/_{\overline{X}}$$

$$\text{Coefficient of median deviation} = MD\Big/_{\widetilde{X}}$$

It is to be noted that coefficient of median deviation is smaller than the coefficient of mean deviation.

vii) The Variance

The variance is defined as the average of the squared deviation from mean. It is denoted by S^2. The formula for the variance may be defined as

$$S^2 = \Sigma\left(X - \overline{X}\right)^2\Big/_n$$

The variance for the grouped data may be defined as

$$S^2 = \Sigma f\left(X - \overline{X}\right)^2\Big/_{\Sigma f}$$

viii) The Standard Deviation

The standard deviation may be defined as the positive square root of the variance. It is denoted by S. The formula for the standard deviation may be defined as

$$S = \sqrt{\Sigma\left(X - \overline{X}\right)^2\Big/_n}$$

The standard deviation for the grouped data may be defined as

$$S = \sqrt{\Sigma f\left(X - \overline{X}\right)^2\Big/_{\Sigma f}}$$

The coefficient of variation may be defined as

$$CV = \left(SD\Big/_{\overline{X}}\right) * 100$$

1.4.2 Probability Distributions

The science of statistics is based on a number of probability distributions used to measure the likelihood of a particular phenomenon. Two types of random variables are considered in the control chart literature, i.e. continuous random variable and the discrete random variable. A continuous random variable is one which can assume each and every possible value in a given interval, for example, height, weight, temperature, etc. Whereas a discrete variable can assume only some specific integer or whole number value as it deals with the counting of the numbers, for example, yes/no, good/defective, go/no go, etc.

Continuous Probability Distributions
Normal Probability Distribution
In the domain of the SPC, mostly we collect the random data on continuous scale (Oakland, 2008). The likelihood of this continuous scale data can be best studied through the normal probability distribution. It is a bell-shaped distribution with the mean at the center and its spread; the width of the curve is reflected by its standard deviation. Due to some nice properties, it is commonly used by the quality control personnel. In SPC the normal probability distribution helps us to determine the probability of falling an observation within the control limits and out of the control limits. If x is the normal random variable, then the probability distribution of x can be defined as

$$f\left(x \mid \mu, \sigma^2\right) = \frac{1}{\sigma\sqrt{2\pi}} e^{-\left(\frac{(x-\mu)^2}{2\sigma^2}\right)} \qquad -\infty < x < \infty$$

The mean of the normal distribution is $\mu(-\infty < \mu < \infty)$, and the variance is $\sigma^2 > 0$. This distribution is commonly used as $x \sim N(\mu, \sigma^2)$ to describe that x is normally distributed with mean μ and variance σ^2.

The cumulative normal distribution may be defined as the probability that the random variable x is less than or equal to some arbitrarily value a, which is

$$P(x \le a) = F(a) = \int_{-\infty}^{a} \frac{1}{\sigma\sqrt{2\pi}} e^{-\left(\frac{(x-\mu)^2}{2\sigma^2}\right)} dx$$

The solution of this integral cannot be computed in closed form. However, by using the transformation of the variable x as

$$z = \frac{x - \mu}{\sigma}$$

with zero mean and unit variance and known as standard normal distribution. When this transformation takes place, the area of the random variable from the non-normal distribution is the same as that of the area of the standard normal distribution. The evaluation can be made independent of μ and σ^2; then this probability can be written as

$$P(x \leq a) = p\left(z \leq \frac{a - \mu}{\sigma}\right) = \Phi\left(\frac{a - \mu}{\sigma}\right)$$

where $\Phi(\cdot)$ is the cumulative distribution function of the standard normal distribution with mean 0 and unit variance (Montgomery, 2009). The areas or the probabilities can be computed using the table of areas under standard normal distribution given in Appendix A.

The graph of the normal distribution is called the normal curve. The normal probability distribution has the following important properties (Figure 1.2):

1) The range of the normal random variable is $-\infty, +\infty$.
2) This distribution is bell shaped and symmetric about mean.
3) The mean, median, and mode of the normal distribution are same.
4) The total area/probability under the normal curve is one.
5) All odd order moments about mean are zero.
6) The normal curve extended to both sides of the mean but never touches the x-axis.
7) The area under the normal curve from

$\mu - \sigma$ to $\mu + \sigma$ is 68.27%

$\mu - 2\sigma$ to $\mu + 2\sigma$ is 95.45%

$\mu - 3\sigma$ to $\mu + 3\sigma$ is 99.73% of almost all the area.

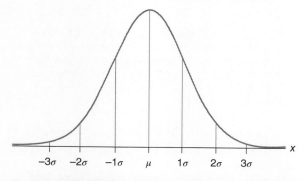

Figure 1.2 Standard normal curve. Source: https://www.google.com/search?q=Standard +normal+curve&safe=strict&rlz=1C1CHBD_enSA905SA905&sxsrf=ALeKk038CFj1c5F9mx FymaEMSjV1xUEkzA:1592774965148&tbm=isch&source=iu&ictx=1&fir=PAgPMxS8f Xpb_M%253A%252CrrsoLwiuhUAKeM%252C_&vet=1&usg=AI4_-kQqYcq7FaH5CrTe620- F-8cvWw6Bg&sa=X&ved=2ahUKEwjdr4SQ7ZPqAhVRPJoKHQTVC50Q_h0wAX oECAcQBg&biw=1280&bih=631#imgrc=PAgPMxS8fXpb_M:

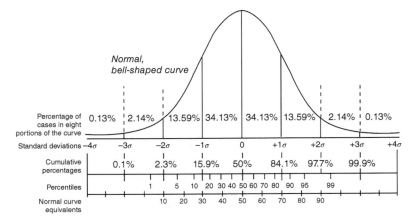

Figure 1.3 Standard normal curve. Source: https://www.google.com/search?q=Standard+normal+curve&safe=strict&rlz=1C1CHBD_enSA905SA905&sxsrf=ALeKk038CFj1c5F9mxFymaEMSjV1xUEkzA:1592774965148&tbm=isch&source=iu&ictx=1&fir=PAgPMxS8fXpb_M%253A%252CrrsoLwiuhUAKeM%252C_&vet=1&usg=AI4_-kQqYcq7FaH5CrTe620-F-8cvWw6Bg&sa=X&ved=2ahUKEwjdr4SQ7ZPqAhVRPJoKHQTVC50Q_h0wAXoECAcQBg&biw=1280&bih=631#imgrc=PAgPMxS8fXpb_M:

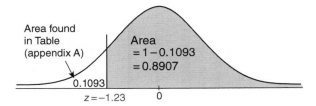

Figure 1.4 Standard normal curve. Source: https://www.google.com/search?q=Standard+normal+curve&safe=strict&rlz=1C1CHBD_enSA905SA905&sxsrf=ALeKk038CFj1c5F9mxFymaEMSjV1xUEkzA:1592774965148&tbm=isch&source=iu&ictx=1&fir=PAgPMxS8fXpb_M%253A%252CrrsoLwiuhUAKeM%252C_&vet=1&usg=AI4_-kQqYcq7FaH5CrTe620-F-8cvWw6Bg&sa=X&ved=2ahUKEwjdr4SQ7ZPqAhVRPJoKHQTVC50Q_h0wAXoECAcQBg&biw=1280&bih=631#imgrc=PAgPMxS8fXpb_M:

Different areas under the normal curve can be calculated as given in Figure 1.3.

To calculate the area $Z \geq -1.23$, we proceed by identifying the area on the normal curve as given in Figure 1.4.

To calculate the area between that $-2.00 \leq Z \leq 1.50$, we proceed by identifying the area on the normal curve as shown in Figure 1.5.

The standard normal distribution has the following important properties:

1) The cumulative area for the Z-score equal to -3.49 is close to 0.
2) The cumulative area for the Z-score increases as the value of Z-score increases.
3) The cumulative area for the Z-score equal to 0 is 0.50.
4) The cumulative area for the Z-score equal to 3.49 is close to 1.

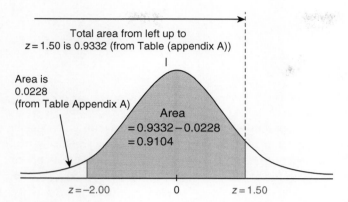

Figure 1.5 Standard normal curve. Source: https://www.google.com/search?q=Standard
+normal+curve&safe=strict&rlz=1C1CHBD_enSA905SA905&sxsrf=ALeKk038CFj1c5F9mx
FymaEMSjV1xUEkzA:1592774965148&tbm=isch&source=iu&ictx=1&fir=PAgPMxS8f
Xpb_M%253A%252CrrsoLwiuhUAKeM%252C_&vet=1&usg=AI4_-kQqYcq7FaH5CrTe620-
F-8cvWw6Bg&sa=X&ved=2ahUKEwjdr4SQ7ZPqAhVRPJoKHQTVC50Q_h0wAX
oECAcQBg&biw=1280&bih=631#imgrc=PAgPMxS8fXpb_M:

Student's t-Distribution

Another most commonly used probability distribution is the Student's t-distribution, which was discovered by the English statistician William Sealy Gossett (1876–1937), when he published his paper with the pseudonym, "Student." This distribution is used for the estimation of the population mean when a sample of v assumed to be normally distributed from that population. If x is a random variable, then the heavy tailed symmetrical t-distribution is defined as

$$f(x \mid v) = \varphi(x; v) = \frac{\Gamma\left(\frac{v+1}{2}\right)}{\sqrt{v\pi}\,\Gamma(v/2)} \left(1 + \frac{x^2}{v}\right)^{-\left(\frac{v+1}{2}\right)} \qquad -\infty < x < \infty, v > 0$$

with mean zero and variance $\dfrac{v}{v-2}$

As v increases, the Student's t-distribution tends to a normal (0, 1) distribution.

Gamma Distribution

Yet another important probability distribution commonly used in the literature of the control charts for non-normal random variables is the Gamma distribution. The probability distribution of the Gamma distribution can be defined as

$$f(x \mid \alpha, \beta) = \frac{\beta^{\alpha}}{\Gamma(\alpha)} x^{\alpha-1} e^{-\beta x} \qquad x \geq 0$$

with the scale parameter $\beta > 0$ and the shape parameter $\alpha > 0$.

The mean and variance of the Gamma distribution are

$$\mu = \frac{\alpha}{\beta}$$

and

$$\sigma^2 = \frac{\alpha}{\beta^2},$$

respectively (Montgomery, 2009).

Discrete Probability Distributions
Binomial Probability Distribution

Let a process consists of a set of n independent trials. Here the term independent means that any outcome is not affected by the previous outcome whether it had occurred or not. Here we define any outcome as either success or failure. Suppose that the probability of success is denoted by p, p belongs to the interval $(0,1)$, and the probability of failure is denoted by $q = 1 - p$, then the binomial probability distribution can be defined as

$$P(X = x) = \frac{n!}{x!(n-x)!} p^x(1-p)^{n-x}$$

where n is the total number of independent trials and x is a binomial random variable ranging from 0 to n. This distribution has only two parameters n and p. The distribution is symmetric when $p = \frac{1}{2}$, the distribution is positively skewed if $p < \frac{1}{2}$, and it is negatively skewed if $p > \frac{1}{2}$.

The binomial probability distribution is the most commonly used distribution in the control chart literature to model the number cases in a sample of n items when the proportion in the population is known. For example, if the proportion of defective item in any mass production unit is 0.12, then find the complete binomial probability distribution for $n = 10$ (Table 1.1).

Poisson Probability Distribution

The Poisson distribution is another important discrete probability distribution used for calculating the characteristics of the control chart, which identifies a given

Table 1.1 Probabilities of number of defective items using binomial distribution.

No. of defective items	Probability	No. of defective items	Probability
0	0.27850098	6	0.00037604
1	0.37977406	7	0.00002930
2	0.23304317	8	0.00000150
3	0.08474297	9	0.00000005
4	0.02022275	10	0.00000000
5	0.00330918	**Total**	1.00000000

The mean and standard deviation of the binomial distribution are $np = 1.2$ and $\sqrt{npq} = 1.264911$, respectively.

Table 1.2 Probabilities of number of defective items using Poisson distribution.

No. of defective items	Probability	No. of defective items	Probability
0	0.110803158	6	0.017448405
1	0.243766948	7	0.005483784
2	0.268143643	8	0.001508041
3	0.196638672	9	0.000368632
4	0.108151269	10	0.000081099
5	0.047586559	11 and more	0.000019789

The mean and standard deviation of this distribution are $\mu = 2.5$ and $\sqrt{\sigma} = 1.58$, respectively.

number of defects per unit; for instance, the number of stones in a piece of glass of the given size or the number of defects in the manufacturing of items, etc. This distribution has only one parameter μ.

The frequency distribution of the Poisson distribution can be defined as

$$P(X = x) = \frac{e^{-\mu} \mu}{x!}$$

where μ = mean number of defects, $\mu > 0$, e = 2.71828... and x = number of occurrences, $x = 0, 1, 2, 3, \ldots$

Suppose that the average number of stones in a glass of a particular shape was 2.5. Find the probabilities of different number of stones in the glass. Evaluating the Poisson distribution in this case, we get the results shown in Table 1.2.

1.5 Types of Control Charts

Broadly speaking, there are two types of control charts, i.e. attribute control charts and variable control charts.

1.5.1 Attribute Control Charts

When articles or units are studied on the basis of qualitative measures as go/no go, yes/no, satisfied/not satisfied, positive/negative, etc., then an attribute control chart is suitable to monitor the unusual changes. These are the charts commonly known as p-chart used for monitoring proportion of nonconforming items in a sample or proportion of defective items in a sample, and np-chart is used to monitor the number of nonconforming items in a sample of size n.

The control limits of these control charts are similar particularly when the sample size is fixed. The limits of p-chart charts may be constructed by using the binomial probability distribution with parameter p. As we know that the mean of the binomial distribution is np and its standard deviation is $\sqrt{np(1-p)}$. Thus

$$\text{UCL} = p + 3\sqrt{p(1-p)/n}$$
$$\text{CL} = p$$
$$\text{LCL} = p - 3\sqrt{p(1-p)/n}$$

And the control limits for the np-chart can be constructed as

$$\text{UCL} = np + 3\sqrt{np(1-p)}$$
$$\text{CL} = np$$
$$\text{LCL} = np - 3\sqrt{np(1-p)}$$

Another attribute control chart that is commonly used in the control chart literature is the c-chart used for monitoring the number of nonconformities in each sample or the number of defects in each sample. The limits of c-chart charts may be constructed by using the Poisson probability distribution with parameter c. As we know that the mean of the Poisson distribution is c and its standard deviation is $\sqrt{c}$. Thus

$$\text{UCL} = c + 3\sqrt{c}$$
$$\text{CL} = c$$
$$\text{LCL} = c - 3\sqrt{c}$$

Yet another attribute control chart commonly used in the literature of the SPC is the μ-chart. It is used to monitor the nonconformity items per unit or the nonconforming items per unit. The distribution of the c-chart and the μ-chart is the same except the scale of the units, which is changed by n in the μ-chart. The control limits of μ-chart can be established as

$$\text{UCL} = \mu + 3\sqrt{\mu/n}$$
$$\text{CL} = \mu$$
$$\text{LCL} = \mu - 3\sqrt{\mu/n}$$

1.5.2   Variable Control Charts

When the characteristics under study are measured on the continuous scale, such as the height, weight, time, and temperature, then such variables are classified as the continuous variables. The commonly used control charts are the average or the dispersion charts or the combination of both the charts. The $\overline{X}$-chart is the most common chart developed with the association of the *R*-chart (also known as $\overline{X}$- and *R*-chart). The development of $\overline{X}$- and *S*-chart or S^2-chart is also very common. These charts have been discussed thoroughly in the latter chapters. In general the distribution of all the variable control charts under study is assumed to be normally distributed. The three-sigma control limits are thus computed with the false alarm rate, the probability that an observation falling outside the control limits is 0.0027. The average run length of the three-sigma control limits can be calculated as $1/0.0027 = 370.3704$, which means that on average the process indicates an out-of-control process after 370 samples if the distribution of observations follows exactly the normal distribution. This is the most commonly used level in evaluating and comparing the process in the control chart literature. The concept of three-sigma control limits is commonly accepted by the industry, which is easy to understand, and applied by the quality control personnel. The variable control limits can be developed using the pre-specified constant values given in Appendix B.

1.6 Meaning of Process Control

The term process control is used to keep the process in control. In statistical terminology a graphical method of presenting a sequence of samples on a chart is called control chart, a tool used for monitoring different changes in the process. A typical control chart has been given in Figure 1.6.

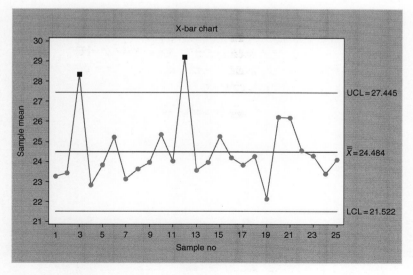

Figure 1.6 Plotting of typical control chart.

References

Ahmad, L., Aslam, M., & Jun, C.-H. (2014). Designing of X-bar control charts based on process capability index using repetitive sampling. *Transactions of the Institute of Measurement and Control, 36*(3), 367–374.

Aslam, M., Srinivasa Rao, G., Ahmad, L., & Jun, C.-H. (2017). A control chart for multivariate Poisson distribution using repetitive sampling. *Journal of Applied Statistics, 44*(1), 123–136. doi:10.1080/02664763.2016.1164837

Garvin, D. (1987). Competing on the eight dimensions of quality. *Harvard Business Review, 65*(6), 101–109.

Montgomery, C. D. (2009). *Introduction to statistical quality control* (6th ed.). New York, NY: Wiley.

Oakland, J. S. (2008). *Statistical process control* (6th ed.). Oxford, UK: Butterworth-Heinemann.

2

Shewhart Type Control Charts for Attributes

In statistical process control (SPC), many situations arise where quality characteristics cannot be easily measured on a numerical scale. However, each inspected item
is usually classified as either conforming or nonconforming to the specifications
of that quality characteristic. The terminology "defective" or "nondefective" is used
frequently to identify these two classifications of a product or the terminology
"conforming" and "nonconforming," respectively. These quality characteristics
are known as attributes. In other word, an **attribute** is a quality characteristic
for which a numerical value is not specified. It is measured on a nominal scale; that
is, it does or does not meet certain guidelines, or it is categorized according to a
scheme of labels. For instance, the taste of a certain dish is labeled as acceptable
or unacceptable or is categorized as exceptional, good, fair, or poor.

A control chart for attributes, on the other hand, is used to monitor characteristics that have discrete values and can be counted. Often they can be evaluated
with a simple yes or no decision. Examples include color, taste, or smell. The monitoring of attributes usually takes less time than that of variables because a variable
needs to be measured (e.g. the bottle of soft drink contains 15.9 oz of liquid). An
attribute requires only a single decision, such as yes or no, good or bad, acceptable
or unacceptable (e.g. the apple is good or rotten, the meat is good or stale, the shoes
have a defect or do not have a defect, the light bulb works or it does not work), or
counting the number of defects (e.g. the number of broken cookies in the box, the
number of dents in the car, the number of barnacles on the bottom of a boat). The
SPC is used to monitor many different types of variables and attributes.

Control charts for attributes were first proposed by Shewhart (1926, 1927) and
have been popularly used in many fields as discussed in Vries and Teneau (2010)
and in Woodall (2006). An early review of control charts designed for attribute data
was given by Woodall (1997). Other studies on reviews and development of attribute control charts included Albers (2009), Chen, Zhou, Chang, and Huang (2008),
Duran and Albin (2009), Famoye (2007), Jahromi, Saghaei, and Ariffin (2012),
Kaminsky, Benneyan, Davis, and Burke (1992), Mohammed, Panesar, Laney,

Introduction to Statistical Process Control, First Edition. Muhammad Aslam, Aamir Saghir,
and Liaquat Ahmad.
© 2021 John Wiley & Sons, Inc. Published 2021 by John Wiley & Sons, Inc.

and Wilson (2003), Saghir, Lin, Abbasi, and Ahmad (2013), Saghir and Lin (2014), Sellers (2012), Szarka and Woodall (2011), Woodall, Tsui, and Tucker (1997), and Woodall (2000). The commonly used attribute control charts are the *p*- and *np*-charts (for binomially distributed processes), the *c*- and *u*-control charts (for Poisson distributed processes), the *g*- and *h*-control charts (for geometric distributed processes), and the generalized chart (for Conway–Maxwell–Poisson [COM–Poisson] distributed processes). All these charts are known as Shewhart attribute charts.

Attribute charts require larger sample sizes than variable charts to ensure adequate protection against a certain level of process changes. Larger sample sizes can be problematic if the measurements are expensive to obtain or the testing is destructive. The choice of sample size for attribute charts is important. It should be large enough to allow nonconformities or nonconforming items to be observed in the sample. For example, if a process has a nonconformance rate of 2.5%, a sample size of 25 is not sufficient because the average number of nonconforming items per sample is only 0.625. Thus, misleading inferences might be made, since no nonconforming items would be observed for many samples. We might erroneously attribute a better nonconformance rate to the process than what actually exists. A sample size of 100 here is sufficient, as the average number of nonconforming items per sample would thus be 2.5. For situations in which summary measures are required, attribute charts are preferred. Information about the output at the plant level is often best described by proportion nonconforming charts or charts on the number of nonconformities. These charts are effective for providing information to upper management. On the other hand, variable charts are more meaningful at the operator or supervisor level because they provide specific clues for remedial actions. What is going to constitute nonconformity should be properly defined. This definition will depend on the product, its functional use, and customer needs. For example, a scratch mark in a machine vise would not be considered nonconformity, whereas the same scratch on a television cabinet would.

2.1 Proportion and Number of Nonconforming Charts

The proportion/fraction of nonconforming is the ratio between the number of nonconforming items in a population and the total number of items in that population. Inspection of several quality characteristics in any item(s) simultaneously may be interest of an inspector. If the item does not conform to standard on one or more of these characteristics, it is classified as **nonconforming or defective**. In many practical situations, a researcher/practitioner may have interest in monitoring proportion of nonconforming or number of nonconforming. For example, this could be the number of nonfunctioning light bulbs, the proportion of broken eggs in a carton, the number of rotten apples, the number of scratches on a tile, or the number of complaints issued. p-Charts are used to measure the fraction nonconforming, while np-charts are used to monitor the number of nonconforming/defective in a product.

The statistical principles underlying the control chart for fraction nonconforming or number nonconforming are based on the binomial distribution. A lot can be classified into two mutually exclusive defectives or nondefective units. Let X denotes the number of defects or the number of nonconforming units in a lot/product from a sample of size n. Then, the random variable X has a binomial distribution with parameters n and p, that is,

$$f(x; n, p) = \binom{n}{x} p^x q^{n-x}, \quad x = 0, 1, 2, ..., n, \quad q = 1 - p. \tag{2.1}$$

The sample proportion of defects/fraction nonconforming can be defined as the ratio between the number of nonconforming and the total units in a sample size n, that is,

$$\hat{p} = \frac{x}{n} \tag{2.2}$$

2.1.1 Proportion of Nonconforming Chart (*p*-Chart)

The proportion/fraction of nonconforming in a lot can be denoted by p. The fraction nonconforming level can be monitored/tested from a batch of lots using the hypotheses:

$H_0 : p = p_0$ (Process is running at the given level of fraction nonconforming)

$H_1 : p \neq p_0$ (Process is not running at the given level of fraction nonconforming)

where p_0 is some target value, which should be as small as possible.

The test statistic is the sample proportion of defectives. The mean and variance of proportion of defectives are $E(\hat{p}) = p_0$ and $\mathrm{SD}(\hat{p}) = \sqrt{\dfrac{p_0(1-p_0)}{n}}$.

The Shewhart three sigma control limits for the p-chart can be defined as

$$\begin{aligned}
\mathrm{LCL} &= p_0 - 3\sqrt{\frac{p_0(1-p_0)}{n}} \\
\mathrm{CL} &= p_0 \\
\mathrm{UCL} &= p_0 + 3\sqrt{\frac{p_0(1-p_0)}{n}}
\end{aligned} \tag{2.3}$$

The parameter or standard is seldom known in practice, and it is measured from the preliminary samples (subgroups) at least 25 as recommend by Montgomery (2013). Assume that we have the observations $X_1, X_2, ..., X_m$ taken over time, where X_i is the number of nonconforming items in a sample whose size is $n_i = n$, $i = 1, 2, ..., m$. The proportion of nonconforming items, p, is estimated by

$$\bar{p} = \frac{\sum\limits_{i=1}^{m} X_i}{\sum\limits_{i=1}^{m} n_i} \tag{2.4}$$

The control limits for monitoring the nonconforming rate with the p-chart when parameter is estimated from Phase I sample are

$$
\begin{aligned}
\text{LCL} &= \bar{p} - 3\sqrt{\frac{\bar{p}(1-\bar{p})}{n}} \\
\text{CL} &= \bar{p} \\
\text{UCL} &= \bar{p} - 3\sqrt{\frac{\bar{p}(1-\bar{p})}{n}}
\end{aligned}
\tag{2.5}
$$

Note that the computed lower control limit (LCL) in Eq. (2.4) is negative, when $\bar{p}$ is small (unless n is very large). In this case, practitioners are unable to use the p-chart to detect process improvements. According to Montgomery (2013), another problem associated with some p-charts is that the upper control limit (UCL) may be so small that the chart signals any time a single nonconforming item is found in a sample. This can result in a higher false alarm rate than desired. One solution to this approach is to choose a value of n large enough such that the probability of at least one nonconforming item that is present in a sample is reasonably large (Montgomery 2013, pp. 298–299). The false alarm rate may also be quite different from desired because of the discreteness of the binomial distribution. Lucas, Davis, and Saniga (2010) recommended adding a value of $1/n$ to the calculated UCL to obtain a smaller false alarm rate.

Acosta-Mejia (1999) showed that the traditional p-chart with control limits from Eq. (2.4) has run length (RL) biased performance. He introduced RL unbiased concept in p-chart and concluded that the new control limits yield an exact (or nearly) RL unbiased chart. For further reading on problems with the p-chart (see Chan, Lai, Xie, & Goh, 2003; Goh, 1987; Goh & Xie, 2003; Ryan & Schwertman, 1997; Xie, Lu, Goh, & Chan, 1999). Xie and Goh (1993) and Schwertman and Ryan (1997) discussed the probability control limits of p-chart with the help of many examples.

Schwertman and Ryan (1999) recommended a dual control chart procedure as a more flexible alternative. Ryan and Schwertman (1997) studied the sensitivity analysis of the *p*-chart, and the fixed probability limits were compared with the traditional three sigma limits using ARL_1. The results using the probability limits were more consistent than the traditional limits. A *p*-chart with adjusted control limits was also introduced using a Poisson approximation to the binomial distribution for $p_0 \leq 0.03$. Other studies on *p*-chart included Xie and Goh (1992), Vaughan (1993), Nelson (1997), Wu, Yeo, and Spedding (2001), Tang and Cheong (2006), and Wang (2009).

The construction of *p*-chart is demonstrated with the help of real-life data set in the following example.

Example 2.1   A production manager at a tire manufacturing plant has inspected the number of defective tires in 20 random samples with 20 observations each. Table 2.1 shows the number of defective tires found in each sample (see https://www.slideshare.net/dchidrewar/control-charts-13158433).

Table 2.1   The number of defective tires in tire manufacturing plant.

Sample no.	No. of defective tires	No. of observations sampled	Fraction defective ($\hat{p}$)
1	3	20	0.15
2	2	20	0.10
3	1	20	0.05
4	2	20	0.10
5	1	20	0.05
6	3	20	0.15
7	3	20	0.15
8	2	20	0.10
9	1	20	0.05
10	2	20	0.10
11	3	20	0.15
12	2	20	0.10
13	2	20	0.10
14	1	20	0.05
15	1	20	0.05
16	2	20	0.10
17	4	20	0.20

(Continued)

Table 2.1 (Continued)

Sample no.	No. of defective tires	No. of observations sampled	Fraction defective ($\hat{p}$)
18	3	20	0.15
19	1	20	0.05
20	1	20	0.05

Check whether the process is working in control by constructing a p-chart.

Solution

From data, $n = 20$, $m = 20$, and $\sum_{i=1}^{m} n_i = 400$, $\sum_{i=1}^{m} X_i = 40$, thus the proportion

of defective tires is $\bar{p} = \dfrac{\sum_{i=1}^{m} X_i}{\sum_{i=1}^{m} n_i} = \dfrac{40}{400} = 0.10$. The control limits of p-chart are

$$\text{LCL} = \bar{p} - 3\sqrt{\frac{\bar{p}(1-\bar{p})}{n}} = -0.101, \text{CL} = \bar{p} = 0.10,$$

$$\text{UCL} = \bar{p} + 3\sqrt{\frac{\bar{p}(1-\bar{p})}{n}} = 0.301$$

Here the LCL is negative and is set to be equal zero because we cannot have negative proportion of defectives. This could be occurs because these are three sigma limits (approximated) not true probability limits. The resulting control chart using the above limits is (Figure 2.1).

Figure 2.1 shows that all the sample fraction defectives ($\hat{p}$) lie within the control limits of the p-chart and declared that the process is in control or working at the desired level of fraction nonconforming/defectives. These initial limits are also called *trial limits*, and one or more points will fall outside these control limits of the p-chart. Then those points are discarded from the data, and control limits will be recalculated based on the remaining data.

Variable Sample Size

There are many reasons why samples vary in size. In processes for which 100% inspection is conducted to estimate the proportion nonconforming, a change in the rate of production may cause the sample size to change. A lack of available inspection personnel or change in the unit cost of inspections is another factor that can influence the sample size. A change in sample size causes the control limits to change, although the central line (CL) remains fixed. As the sample size increases, the control limits become narrower. As stated previously, the sample size is also influenced by the existing average process quality level. For a given process

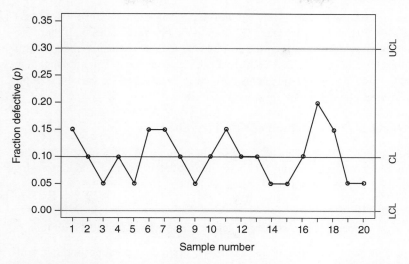

Figure 2.1 The *p*-chart for the number of defective tires in tire manufacturing plant.

proportion nonconforming, the sample size should be chosen carefully so that there is ample opportunity for nonconforming items to be represented. Thus, changes in the quality level of the process may require a change in the sample size.

Control limits can be constructed for individual samples. If no standard is given and the sample average proportion nonconforming is p, the control limits for sample i with size n_i are

$$\text{LCL} = \bar{p} - 3\sqrt{\frac{\bar{p}(1-\bar{p})}{n_i}}$$

$$\text{UCL} = \bar{p} - 3\sqrt{\frac{\bar{p}(1-\bar{p})}{n_i}}.$$

Improved *p*-Chart

The classic control charts for attribute data (*p*-charts, *u*-charts, etc.) are based on assumptions about the underlying distribution of their data (binomial or Poisson). Inherent in those assumptions is the further assumption that the "parameter" (mean) of the distribution is constant over time. In real applications, this is not always true (some days it rains, and some days it does not). This is especially noticeable when the subgroup sizes are very large.

Laney (2002) introduced a new tool, the $\acute{p}$-chart, which solves that problem. In fact, it is a universal technique that is applicable to know whether the parameter is stable or not. The $\acute{p}$-chart does not have to choose between intrasubgroup variation (as in the *p*-chart) or inter-subgroup variation (as in the *X*-chart). It uses all the variation in the data. If there is any batch-to-batch variation, its control limits are appropriately farther away from the CL than in a p-chart. In addition, if there is a variation in subgroup sizes, its control limits will vary, unlike the X-chart. In this procedure, the concepts of the X-chart and the z-chart are implemented together. Convert the p-values to z-scores (thus correcting in advance for variable sample sizes) and then plot the z's in an individual's chart.

2.1.2 Number of Nonconforming Chart (np-Chart)

It is also possible to base a control chart on the number of nonconforming rather than the fraction nonconforming. This is often called as number nonconforming (np) control chart to test nonconforming level using the hypotheses:

$H_0 : np = np_0$ (Process is working at desired mean number of nonconforming)

$H_1 : np \neq np_0$ (Process is not working at desired mean number of nonconforming)

where p_0 is some target value, which should be as small as possible, and n is sample size.

The control limits for np-chart, which monitor the number of nonconforming items, are

$$\begin{aligned}
\text{LCL} &= n\bar{p} - 3\sqrt{n\bar{p}(1-\bar{p})} \\
\text{CL} &= n\bar{p} \\
\text{UCL} &= n\bar{p} - 3\sqrt{n\bar{p}(1-\bar{p})}
\end{aligned} \qquad (2.6)$$

2.1.3 Performance Evaluation Measures

In control chart applications, various performance criteria are used for determining the best method over a range of shifts. The operating characteristic (OC) curve is one of these measures, which is a graphical display of the probability of incorrectly accepting the hypothesis of statistical control (i.e. a type II or β error) against the range of shifts. With reference to the p- or np-chart, the OC curve provides a measure of the sensitivity of the control chart, that is, its ability to detect a shift in the process fraction/number nonconforming from the nominal value p_0 to some other value p_1, where $p_1 \neq p_0$. The probability of type II error for the fraction nonconforming control chart may be computed from

$$\begin{aligned}
\beta &= \Pr(\text{incorrectly accepting the hypothesis of statistical control}) \\
\beta &= \Pr(\hat{p} < \text{UCL}|p_1) - \Pr(\hat{p} < \text{LCL}|p_1) \\
\beta &= \Pr(X < n\text{UCL}|p_1) - \Pr(X < n\text{LCL}|p_1)
\end{aligned} \qquad (2.7)$$

where X is a binomial random variable having probability density function (p.d.f.) given in Eq. (2.1). The value β defined in Eq. (2.7) can be obtained for the given values of n and p_1 using the cumulative binomial distribution function. Note that when the LCL is negative, the second term on the right-hand side of Eq. (2.7) should be dropped.

Another criterion for determining the performance evaluation of any chart over a range of shift is average run length (ARL). The ARL is defined as the expected number of plotted points on a control chart before the first signal appears, i.e.

$$\text{ARL} = 1/\text{Pr}(\text{sample point plots out of control})$$

Thus, if the process is in control, the ARL is

$$\text{ARL}_0 = 1/\alpha$$

and if it is out of control, then

$$\text{ARL}_1 = 1/1 - \beta$$

This is an appropriate measure when using equal sample sizes, and sampling intervals for all methods are compared. The probabilities (α, β) can be obtained from the cumulative density function of binomial distribution for the given values of n and p_1.

We consider the above example to demonstrate the calculation of performance measures of the p-chart. As $n = 20$, LCL $= 0$, UCL $= 0.301$, and $p = 0.10$, we use this information, and β is calculated as

$$\beta = \text{Pr}(X < 20 * 0.301 | p_1) - \text{Pr}(X < 20 * 0 | p_1)$$
$$\beta = \text{Pr}(X < 6.02 | p_1) - \text{Pr}(X < 0 | p_1)$$

Since X must be integer in binomial distribution, we may write as (Table 2.2 and Figure 2.2)

$$\beta = \text{Pr}(X \le 6 | p_1) - \text{Pr}(X \le 0 | p_1)$$

Table 2.2 Calculations of β function for p-chart with $n = 20$, LCL = 0, and UCL = 0.301.

p	$\text{Pr}(X \le 6 | p_1)$	$\text{Pr}(X \le 0 | p_1)$	$\beta = \text{Pr}(X \le 6 | p_1) - \text{Pr}(X \le 0 | p_1)$
0.06	0.9710	0.0453	0.9257
0.08	0.8981	0.0154	0.8826
0.10	0.7702	0.0051	0.7650
0.12	0.6065	0.0016	0.6048
0.14	0.4383	0.0005	0.4378
0.16	0.2919	0.0001	0.2917
0.18	0.1800	0.0000	0.1800
0.20	0.1034	0.0000	0.1033
0.25	0.0193	0.0000	0.0193
0.30	0.0024	0.0000	0.0024

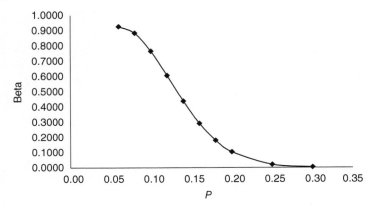

Figure 2.2 The OC curve for tire manufacturing plant process with LCL = 0, UCL = 0.301, and *n* = 20.

2.2 Number of Nonconformities and Average Nonconformity Charts

As defined in previous section, a nonconforming item is a unit of product that does not satisfy one or more of the specifications for that product. Thus, a nonconforming item may have at least one nonconformity unit. In many processes, it is quite possible for a unit to contain several nonconformities and not be classified as nonconforming. As an example, suppose we are manufacturing personal computers. Each unit could have one or more very minor flaws in the cabinet finish, and since these flaws do not seriously affect the unit's functional operation, it could be classified as conforming. However, if there are too many of these flaws, the personal computer should be classified as nonconforming, since the flaws would be very noticeable to the customer and might affect the sale of the unit. There are many practical situations in which we prefer to work directly with the number of defects or nonconformities rather than the fraction nonconforming. These include the number of defective welds in 100 m of oil pipeline, the number of broken rivets in an aircraft wing, the number of functional defects in an electronic logic device, and the number of errors on a document (see Montgomery, 2013).

The number of defects and average number of defects follow the Poisson distribution. Therefore, it is possible to construct the number of defect (*c*-) chart or the average number of defect (*u*-) chart using the Poisson distribution and its cumulative density function. Let *X* denotes the number of defects or the number of nonconformities in an inspected unit of product. Then, the random variable *X* has a Poisson distribution having p.d.f.:

$$f(x; c) = \frac{e^{-c}c^x}{x!}, \quad x = 0, 1, 2, \dots \tag{2.8}$$

where c is the parameter of Poisson distribution and denotes the average nonconformities/defects. The sample estimator of c based on m preliminary samples of size n is

$$\bar{c} = \sum_{i=i}^{m} C_i/m \tag{2.9}$$

2.2.1 Number of Nonconformities (*c*-) Chart

The c-chart counts the actual number of defects. For example, we can count the number of complaints from customers in a month, the number of bacteria on a petri dish, or the number of barnacles on the bottom of a boat. However, we cannot compute the proportion of complaints from customers, the proportion of bacteria on a petri dish, or the proportion of barnacles on the bottom of a boat. We test the following hypotheses:

$H_0 : c = c_0$ (Process working at given number of nonconformities per lot)

$H_1 : c \neq c_0$ (Process is not working at desired number of nonconformities per lot)

For the c-chart, the data consist of C_i, $i = 1, 2, \dots, m$, where C_i is the number of nonconformities found on the ith sample of n units of product. The values of C_i are plotted against the three sigma control limits defined as

$$\begin{aligned} \text{LCL} &= \bar{c} - 3\sqrt{\bar{c}} \\ \text{CL} &= \bar{c} \\ \text{UCL} &= \bar{c} + 3\sqrt{\bar{c}} \end{aligned} \tag{2.10}$$

where $\bar{c} = \sum_{i=i}^{m} C_i/m$.

According to Montgomery (2013), defect or nonconformity data are always more informative than fraction nonconforming, because there will usually be several different types of nonconformities. By analyzing the nonconformities by type, we can often gain considerable insight into their cause. This can be of considerable assistance in developing the out-of-control action plans that must accompany control charts.

Example 2.2 Table 2.3 data represent the number of nonconformities per 1000 m in telephone cable (see Montgomery, 2013). From analysis of these data, would you conclude that the process is in statistical control? What control procedure would you recommend for future production?

Table 2.3 The number of nonconformities per 1000 m in telephone cable.

Sample no.	No. of nonconformities	Sample no.	No. of nonconformities
1	1	12	6
2	1	13	9
3	3	14	11
4	7	15	15
5	8	16	8
6	10	17	3
7	5	18	6
8	13	19	7
9	0	20	4
10	19	21	9
11	24	22	20

Solution

From data, $\bar{c} = 8.59$. Therefore, the three sigma control limits of c-chart are

$$\text{LCL} = \bar{c} - 3\sqrt{\bar{c}} = -0.2026, \text{CL} = \bar{c} = 8.59, \text{and UCL} = \bar{c} + 3\sqrt{\bar{c}} = 17.383$$

here the LCL is negative and is set to be equal zero because there could not be negative defects in a lot. This could occur because we used 3σ limits that assumed normal distribution assumption. The c-chart is plotted in Figure 2.3.

Figure 2.3 shows that the process is not in statistical control as three subgroups (10th, 11th, and 22nd) exceed the UCL. These three subgroups are discarded, and the control limits are recalculated as

$$\text{LCL} = \bar{c} - 3\sqrt{\bar{c}} = -1.2818 = 0, \text{CL} = \bar{c} = 6.17, \text{and UCL} = \bar{c} + 3\sqrt{\bar{c}} = 13.622$$

The c-chart is again constructed and plotted in Figure 2.4.

Now, the sample subgroup 15 exceeds the UCL; after deleting this point, the control limits should be constructed and that limits should be used in Phase II process monitoring.

2.2.2 Average Nonconformities (u-) Chart

In c-chart the sample size is chosen exactly equal to one inspection unit. However, there is no reason why the sample size must be restricted to one inspection unit. If the quality is measured in terms of the average number of defects per unit with

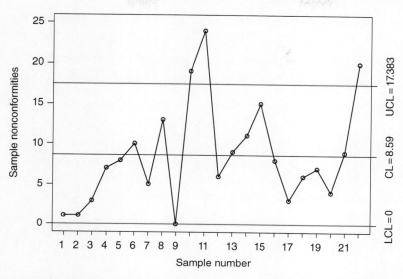

Figure 2.3 The *c*-chart for the number of nonconformities per 1000 m in telephone cable.

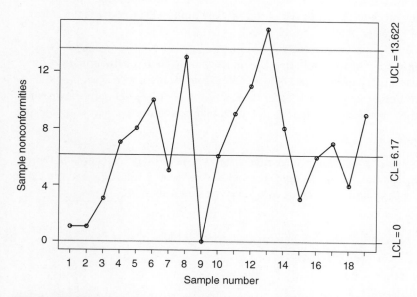

Figure 2.4 The *c*-chart for the number of nonconformities per 1000 m in telephone cable after 10th, 11th, and 22nd batches discarded.

inspection unit more than 1 (and not the actual number of defects per unit or the actual number of defects per sample batch), then alternatively a u-chart is used:

$H_0 : u = u_0$ (Process working at given average number of defects per unit)

$H_1 : u \neq u_0$ (Process not working at given average number of defects per unit)

here, u denotes the average number of defects per unit and defined as

$$u = \frac{x}{n} \tag{2.11}$$

where x denotes the total nonconformities in a sample of n inspection units.

As x is a Poisson random variable, resultantly, the control limits for the average number of nonconformities per unit (u-) chart could be defined as follows:

$$\begin{aligned} \text{LCL} &= \bar{u} - 3\sqrt{\bar{u}} \\ \text{CL} &= \bar{u} \\ \text{UCL} &= \bar{u} + 3\sqrt{\bar{u}} \end{aligned} \tag{2.12}$$

where $\bar{u}$ represents the observed average number of nonconformities per unit in a preliminary set of data. These control limits would be considered as trial control limits, and the U_i is plotted against the control limits, and process is declared in control if all the observed nonconformities per unit lie within the limits. Otherwise, the process is declared out of control, and limits will be revised after discarding the outsided points from the data.

Example 2.3 A paper mill uses a control chart to monitor the imperfection in finished rolls of paper (Table 2.4) (data taken from Montgomery, 2013, p. 339). Production output is inspected for 20 days, and the resulting data are shown here. Does the process is in statistical control state by assuming average sample size?

Solution

As number of rolls produced (sample size) vary day by day (batches/subgroups), we may use the average sample size, i.e. 20. We have $\bar{u} = 0.7007$, and the three sigma control limits of u-chart using average sample size are

$$\text{LCL} = \bar{u} - 3\sqrt{\frac{\bar{u}}{n}} = 0.7007 - 3\sqrt{\frac{0.7007}{20}} = 0.1468,$$

$$\text{CL} = \bar{u} = 0.7007,$$

$$\text{UCL} = \bar{u} + 3\sqrt{\frac{\bar{u}}{n}} = 0.7007 + 3\sqrt{\frac{0.7007}{20}} = 1.2547.$$

Table 2.4 The imperfection in finished rolls of paper.

Day	No. of rolls produced	Total no. of imperfection	Day	No. of rolls produced	Total no. of imperfection
1	18	12	11	18	18
2	18	14	12	18	14
3	24	20	13	18	9
4	22	18	14	20	10
5	22	15	15	20	14
6	22	12	16	20	13
7	20	11	17	24	16
8	20	15	18	24	18
9	20	12	19	22	20
10	20	10	20	21	17

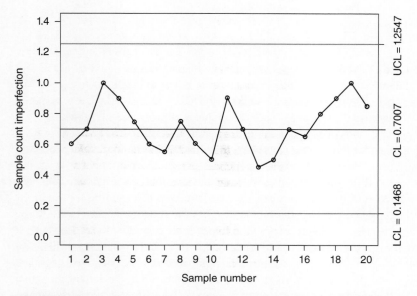

Figure 2.5 The u-chart for the number of imperfection in paper mill production.

The u-chart is plotted using the sample count imperfection against the control limits in Figure 2.5.

All the points fall within the control limits, and it can be concluded that the process is in statistical control state.

2.2.3 The Performance Evaluation Measure

The performance of the c-chart and u-chart can be evaluated using different measures. The OC function of both the charts can be obtained from the Poisson distribution as given in many books including Montgomery (2013). The probability of type II error (OC function) for the c-chart can be determined as

$$\beta = \text{Pr(incorrectly accepting the hypothesis of statistical control)}$$
$$\beta = \text{Pr}(\text{LCL} \leq x < \text{UCL}|c_1)$$

$$\beta = \sum\nolimits_{x=\text{LCL}}^{\text{UCL}} \frac{e^{-c_1}c_1^{x}}{x!} \tag{2.13}$$

where X follows a Poisson distribution with p.d.f. given in Eq. (2.8). The value β defined in Eq. (2.7) can be obtained for the given values of c_1.

Similarly, the OC function for the u-chart can be determined by the following equations:

$$\beta = \text{Pr}(\text{LCL} \leq U < \text{UCL}|u_1)$$
$$\beta = \text{Pr}(n\text{LCL} \leq x < n\text{UCL}|u_1)$$

$$\beta = \sum\nolimits_{x=\text{LCL}}^{\text{UCL}} \frac{e^{-(nu_1)}(nu_1)^{x}}{x!} \tag{2.14}$$

where $(n\text{LCL})$ denotes the smallest integer greater than or equal to $n\text{LCL}$ and $(n\text{UCL})$ denotes the largest integer less than or equal to $n\text{UCL}$.

Using Eq. (2.13), the β values of c-chart with UCL $= 33.22$ and LCL $= 6.48$ (taken from Montgomery, 2013, p. 233) are calculated and presented in Table 2.5.

The control limits of c- and u-charts defined in Eqs. (2.10) and (2.12) have been widely applied for monitoring count data in industrial and nonindustrial processes. However, using these limits always experiences an excessive amount of false alarms, since these control limits are defined by impractical normal assumption for the right skewed Poisson distribution. Probability-based methods determining control limits were discussed by many authors including Ryan and Schwertman (1997). Many other approaches have been proposed to improve the c- and u-charts. However, it is possible to categorize them into three major groups of approaches, which are the following: (i) the transforming data approach, (ii) the standardizing data approach, and (iii) the optimizing control limits approach. Aebtarm and Bouguila (2001) provided a detailed review on the empirical evaluation of these approaches. Braun (1999) derived the RL distribution for the c- and p-charts when the control limits are estimated. ARL and standard deviation of run length (SDRL) were used as performance measures to evaluate these two charts. Chakraborti and Human (2008) determined the exact RL distribution for the c-chart in Phase II applications. Expressions for various chart performance characteristics, such as the ARL, SDRL, and the median run length (MDRL), were also obtained.

Table 2.5 The β values of c-chart with UCL = 33.22 and LCL = 6.48.

c_1	$Pr\{x \leq 33\}$	$Pr\{x \leq 6\}$	β
1	1.000	0.999	0.001
3	1.000	0.966	0.034
5	1.000	0.762	0.238
7	1.000	0.450	0.550
10	1.000	0.130	0.870
15	0.999	0.008	0.991
20	0.997	0.000	0.997
25	0.950	0.000	0.950
30	0.744	0.000	0.744
33	0.546	0.000	0.546
35	0.410	0.000	0.410
40	0.151	0.000	0.151
45	0.038	0.000	0.038

Dealing with Low Defect Levels

When defect levels or **count rates** in a process become very low – say, under 1000 occurrences per million – there will be very long periods of time between the occurrences of a nonconforming unit. In these situations, many samples will have zero defects, and a control chart with the statistic consistently plotting at zero will be relatively uninformative. Thus, conventional c- and u-charts become ineffective as count rates are driven into the low parts per million range.

One way to deal with this problem is to adopt a **time between occurrence control chart**, which charts a new variable: the time between the successive occurrences of the count. The time-between-events control chart has been very effective as a process-control procedure for processes with low defect levels. Suppose that defects or counts or "events" of interest occur according to a Poisson distribution. Then the probability distribution of the time between events is the exponential distribution. Therefore, constructing a time-between-events control chart is essentially equivalent to control charting an exponentially distributed variable. However, the exponential distribution is highly skewed, and as a result, the corresponding control chart would be very asymmetric. Such a control chart would certainly look unusual and might present some difficulties in interpretation for operating personnel.

2.3 Control Charts for Over-Dispersed Data

2.3.1 Dispersion of Counts Data

Traditionally, it is assumed that the quality characteristics of counts process follow either a binomial distribution or a Poisson distribution. The Poisson distribution is a popular distribution used to describe count information, from which control charts involving count data have been established in the literature (e.g. Montgomery, 2013; Sellers, 2012). The Poisson distribution depends on a single parameter, λ, which is the mean as well as the variance (**equi-dispersion**). In many count data sets, however, the equi-dispersion assumption is limiting, and a generalized control chart to monitor over- or under-dispersed data is required. In such situations, one plausible approach is to use the binomial (or Bernoulli) distribution, when the mean is greater than variance (a case of **under-dispersion**), or the negative binomial (or geometric) distribution, when the mean is smaller than variance (a case of **over-dispersion**).

In many attribute processes, the ratio of the sample variance to the sample mean is usually larger than 1. This type of data is called over-dispersed data and is common in many fields (e.g. From, 2004; Luceno, 2005; Shankar, Milton, & Mannering, 1997). Focusing on data sets that illustrate over-dispersion, the application of Poisson model increases the false alarm rate (falsely detected as out of control when, in fact, points are in control). To monitor such data, a reasonable approach is to use the negative binomial distribution (which has the geometric distribution as a special case) as discussed by Albers (2009) and Zhang, Peng, Schuh, Megahed, and Woodall (2013).

2.3.2 *g*-Chart and *h*-Chart

Kaminsky et al. (1992) developed two new control charts, namely, the control charts to monitor the total number of events (*g*-chart) and the average number of events per unit (*h*-chart), via the geometric distribution. They used the shifted geometric distribution defined as

$$f(x;p) = p(1-p)^{x-a}, \quad x = a, a+1, a+2, \ldots. \tag{2.15}$$

where a is the known minimum possible number of events. Suppose that the data from the process is available as a subgroup of size n, say, $x_1, x_2, \ldots, x_n$ that are independently and identically distributed observations from a geometric distribution. The following two statistics can be measured from the given data, and control charts could be constructed for monitoring the total number of events

$$T = \sum_{i=1}^{n} x_i$$

and the average number of events

$$\overline{T} = \frac{\sum_{i=1}^{n} x_i}{n},$$

respectively. It is well known in the literature that the sum of independently and identically distributed geometric random variables is a negative binomial random variable. This would be useful information in constructing OC curves or calculating ARLs for g- and h-control charts. The g- (total number of events) chart can be constructed by plotting the T_i against the k-sigma control limits defined as

$$
\begin{aligned}
\text{LCL} &= n\left(\frac{1-p}{p} + a\right) - k\sqrt{\frac{n(1-p)}{p^2}} \\
\text{CL} &= n\left(\frac{1-p}{p} + a\right) \\
\text{UCL} &= n\left(\frac{1-p}{p} + a\right) + k\sqrt{\frac{n(1-p)}{p^2}}
\end{aligned}
\tag{2.16}
$$

Similarly, the h- (average number of events) chart can be constructed by plotting the $\overline{T}_i$ against the k-sigma control limits defined as

$$
\begin{aligned}
\text{LCL} &= \left(\frac{1-p}{p} + a\right) - k\sqrt{\frac{(1-p)}{np^2}} \\
\text{CL} &= \left(\frac{1-p}{p} + a\right) \\
\text{UCL} &= \left(\frac{1-p}{p} + a\right) + k\sqrt{\frac{(1-p)}{np^2}}
\end{aligned}
\tag{2.17}
$$

where k is the control charting constant and usually assumes 3 in Shewhart chart. However, the exact value of k can be determined for the fixed value of n and a.

Xie and Goh (1993) studied the use of probability limits instead of the traditional k-sigma limits of the g-chart and h-chart. They discussed the problems associated with the use of three sigma control limits for a number of counts charts. They used negative binomial distribution in the construction and evaluation of probability control limits.

If the subgroup of size n is drawn from geometric distribution defined in Eq. (2.15), then the total number of counts in the subgroup, $Z = X_1 + X_2 + \ldots + X_n$, has a negative binomial distribution or Pascal distribution, that is,

$$P(Z = z) = \binom{n + z - 1}{n - 1} p^n (1 - p)^z, \quad z = 0, 1, 2, \ldots$$

Assuming that p is known and n is fixed, control limits can easily be obtained based on these results. The true probability limits of total number of events (T) chart, namely, g-chart, given by Xie and Goh (1993), are

$$\text{LCL} = \sum_{t=0}^{\text{LCL}} \binom{n + t - 1}{n - 1} p^n (1 - p)^t \leq \frac{\alpha}{2}$$

$$\text{CL} = n\left(\frac{1 - p}{p}\right) \tag{2.18}$$

$$\text{UCL} = \sum_{t = \text{UCL}}^{\infty} \binom{n + t - 1}{n - 1} p^n (1 - p)^t \leq \frac{\alpha}{2}$$

and the true probability limits of average number of events ($\overline{T}$) chart, namely, h-chart, are

$$\text{LCL} = \sum_{t=0}^{\text{LCL}} \binom{n + t - 1}{n - 1} p^n (1 - p)^t \leq \frac{\alpha}{2}$$

$$\text{CL} = \left(\frac{1 - p}{p}\right) \tag{2.19}$$

$$\text{UCL} = \sum_{t = \text{UCL}}^{\infty} \binom{n + t - 1}{n - 1} p^n (1 - p)^t \leq \frac{\alpha}{2}$$

These control limits can be calculated using the negative binomial probability density function for the given value of false alarm rate α. If $n = 1$, which is a common case when, for example, the inspection is carried out automatically, the control limits can easily be derived by using

$$\sum_{t=0}^{t_0} \binom{n + t - 1}{n - 1} p^n (1 - p)^t = 1 - (1 - p)^{t_0}$$

and the LCL and UCL can be given by

Table 2.6 The average run length of the *g*-chart for $p_0 = 0.20$ and $n = 5$.

p	ARL_α	ARL_k	p	ARL_α	ARL_k
0.10	5.329	2.897	0.26	824.8	2,381
0.12	11.97	5.127	0.28	579.1	7,326
0.14	29.95	9.962	0.30	411.4	23,855
0.16	81.84	21.00	0.32	298.0	82,225
0.18	235.9	47.64	0.34	220.1	300,240
0.20	**635.7**	**115.6**	0.36	165.4	$>10^6$
0.22	1141	298.6	0.38	126.2	$>10^6$
0.24	1119	819.1	0.40	97.67	$>10^6$

$$LCL = \frac{\ln(1 - \alpha/2)}{\ln(1 - p)}$$

$$UCL = \frac{\ln(\alpha/2)}{\ln(1 - p)}.$$

Because in the case $n = 1$, negative binomial distribution reduces to geometric distribution, and these control limits are the same as those found in Xie and Goh (1993).

It can also be pointed out here that by using the exact probability limits, the requirement for LCL to be greater than zero is

$$P(Z = 0) = p^n < \alpha/2.$$

This is easily satisfied in practice. For example, when α is 0.0027 and $n = 5$, p has to be greater than 0.267 for LCL to fall below zero. For $p < 0.267$, process changes can be detected even with simple one-point out-of-control rule.

Table 2.6 gives the ARL of the g-chart for (i) true probability limits (ARL_α) and (ii) k-sigma limits (ARL_k) as determined by Xie and Goh (1993).

Example 2.4   Table 2.7 shows that the number of orders on each truck was of interest. The five trucks were considered a shipment. Does the number of orders on each truck process is in statistical control?

Solution

As Kaminsky et al. (1992) showed that data follows the geometric distribution with location parameter 1. Therefore, we can construct the total number of events (*g*-)

Table 2.7 The number of orders on each truck.

Ship no.	Total no. of orders on five trucks	Ship no.	Total no. of orders on five trucks
1	133	11	23
2	95	12	29
3	83	13	24
4	94	14	19
5	50	15	23
6	63	16	23
7	51	17	9
8	54	18	9
9	32	19	15
10	46	20	11

Source: Data is taken from Kaminsky et al. (1992).

chart for the given data. We have $n = 5$ and $\bar{t} = 44.30$, and the three sigma control limits of g-chart are as follows:

$$LCL = 0, CL = 44.30, \text{ and } UCL = 100.28.$$

The g-chart is plotted using the total number of orders on five trucks against the control limits in Figure 2.6.

The first shipment total numbers of orders on five trucks exceed the UCL, and the g-chart limits should be recalculated after discarding the first shipment.

2.4 Generalized and Flexible Control Charts for Dispersed Data

Control charts for under- or over-dispersed count data can be constructed based on either the binomial distribution or the negative binomial distribution (which has the geometric distribution as a special case) as given in previous sections. However, the use of a generalized distribution, which possesses both properties of under-dispersion and over-dispersion, is preferred. This section provides an overview of control charts, which are based on such generalized distributions.

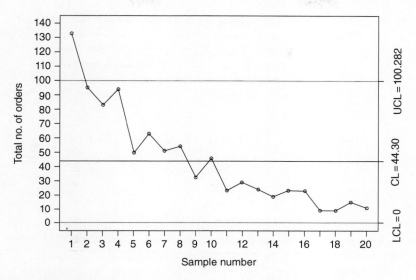

Figure 2.6 The *g*-chart for the total number of orders on five trucks.

2.4.1 The *gc*- and the *gu*-Charts

Famoye (2007) used the shifted (or zero-truncated) generalized Poisson distribution (SGPD) to describe the occurrence of events in production processes. The SGPD is defined as

$$f(x; \theta, \lambda) = \frac{\theta(\theta + \lambda x)^{x-1} e^{-\lambda x}}{(e^\theta - 1)x!}, \quad x = 1, 2, 3... \tag{2.20}$$

where $\theta > 0$ is a location parameter and $0 \leq \lambda < 1$ is a dispersion parameter of the SGPD. The mean and variance of SGPD are, respectively, defined as

$$E(X) = \mu = \theta(1 - \lambda)^{-1}(1 - e^{-\theta})^{-1} \text{ and}$$

$$\text{Var}(X) = \sigma^2 = \left[\theta(1 - \lambda)^{-3} + \theta^2(1 - \lambda)^{-2}\right](1 - e^{-\theta})^{-1} - (E(X))^2$$

Let the subgroup $X_1, X_2, ..., X_n$ be a random sample of size n from the process following SGPD. There following two statistics can be measured from the given data: the total number of events $Y = \sum_{i=1}^{n} x_i$ and the average number of events $\overline{Y} = \frac{Y}{n}$, respectively. Two control charts can be constructed based on these two statistics. One monitors the total number of events, namely, *gc*-chart, and the other monitors the average number of events, namely, *gu*-chart, respectively.

Table 2.8 Control limits for total number of events and average number of events based on the SGPD, respectively.

	Total number of events	Average number of events
LCL	$n\left(\theta\,(1-\lambda)^{-1}\left(1-\mathrm{e}^{-\theta}\right)^{-1}\right) - k\sqrt{n\cdot\sigma^2}$	$\theta\,(1-\lambda)^{-1}\left(1-\mathrm{e}^{-\theta}\right)^{-1} - k\sqrt{\frac{\sigma^2}{n}}$
CL	$n(\theta\,(1-\lambda)^{-1}(1-\mathrm{e}^{-\theta})^{-1})$	$\theta\,(1-\lambda)^{-1}(1-\mathrm{e}^{-\theta})^{-1}$
UCL	$n\left(\theta\,(1-\lambda)^{-1}\left(1-\mathrm{e}^{-\theta}\right)^{-1}\right) - k\sqrt{n\cdot\sigma^2}$	$\theta\,(1-\lambda)^{-1}\left(1-\mathrm{e}^{-\theta}\right)^{-1} + k\sqrt{\frac{\sigma^2}{n}}$

The k-sigma control limits proposed by Famoye (2007) using the SGPD are given in Table 2.8.

He developed control charts for the total number of events (namely, gc-chart) and for the average number of events (namely, gu-chart). Using numerical example analysis, he concluded that the SGPD can better describe an over-dispersed or an under-dispersed data set than the geometric distribution. This model, however, gets truncated under certain conditions regarding the dispersion parameter and thus is not a true probability model.

2.4.2 Control Chart Based on Generalized Poisson Distribution

He, Xie, Goh, and Tsui (2006) proposed a geometric control chart based on generalized Poisson distribution (GPD) for monitoring over-dispersed data. The GPD is a more direct generalization of the Poisson distribution. It has a very versatile nature and similar complexity as the zero-inflated Poisson distribution. It can be noted that both distributions can be used, but the model selection and fitting are different issues from what are considered by He et al. (2006), which focus on general Poisson distribution. The GPD has two parameters (θ, λ) with probability mass function (p.m.f.) given as (He et al., 2006):

$$f(x; \theta, \lambda) = \frac{\theta(\theta + \lambda x)^{x-1}\mathrm{e}^{-\theta-\lambda x}}{x!}, \quad x = 0, 1, 2, 3, \dots \tag{2.21}$$

Theoretically, the parameter λ can also take negative values. However, since we are more interested in the study of over-dispersed data, which corresponds to the situation when λ is positive, we assume $\lambda \geq 0$ here. The mean and variance of the GPD are

$$\mu = E(X) = \theta(1-\lambda)^{-1} \text{ and } \sigma^2 = \theta(1-\lambda)^{-3}$$

respectively. It should be pointed out that the GPD model is easy to use since there are closed-form expressions for both mean and variance, and moment estimates can be easily calculated. On the other hand, the maximum likelihood estimates can also be obtained in a straightforward manner. Consider a set of independent observations $\{X_1, X_2, ..., X_n\}$ with sample size n. Then the log-likelihood function is

$$\ln L(\theta, \lambda) = n \ln \theta + \sum_{i=1}^{n} (x_i - 1) \ln (\theta + x_i \lambda) - n\theta - n\bar{x}\lambda - \sum_{i=1}^{n} \ln x_i!,$$

(2.22)

where $\bar{x} = \sum_{i=1}^{n} x_i/n$. Taking derivatives with respect to the parameters, the maximum likelihood estimators $(\hat{\theta}, \hat{\lambda})$ can be obtained by solving Eq. (2.22).

In the GPD model, two parameters (θ, λ) are involved. The process is stable only when both parameters are not changed. Using a single chart to monitor the process is insufficient, because it is difficult to tell which parameter should be checked when there is an out-of-control alarm. The information about which parameter has shifted is very helpful when assignable causes have to be found. For example, this information can guide the process engineer to find out the cause by checking on relevant factors relating to the specific parameter in question. For this reason, it is preferable to use one chart for each parameter. This idea is quite similar to using the traditional X-bar chart and R chart together to monitor normally distributed variables.

Process Monitoring

When process characteristics can be modeled with the GPD, exact probability control limits can be obtained. However, the LCL for the GPD model may not exist. This is because the probability of zero defect could be larger than the desired type I error probability. This phenomenon is common for attribute control charts. The UCL will be studied here, and a combined procedure for process improvement monitoring is discussed, which is important in SPC.

The UCL for a control chart based on the number of defects can be obtained as the smallest integer solution of the following equation:

$$P \text{ (More than UCL}_c \text{ defects in a sample point)} \leq \alpha_L$$

where α_L is the predetermined false alarm probability for the UCL n_U.

To study the sensitivity of the monitoring procedure, both the OC function and ARL are examined. The GPD model has two parameters, and the two parameters could have different impacts on the alarm probability. The OC function, expressed by the type II error probability β, is a measure of the inability of a control chart to detect process shifts. For the GPD model, assuming that the parameters shift from (θ_0, λ_0) to (θ, λ), the general formula of the type II error probability is

$$\beta = P(X \leq \text{UCL}(\theta_0, \lambda_0)|\theta, \lambda, \alpha_L) = \sum_{k=0}^{\text{UCL}} \frac{\theta(\theta + \lambda k)^{k-1} e^{-\theta - \lambda k}}{k!}$$

where X is the number of defects in a product unit, which is a GPD random variable with parameters θ and λ, and α_L is the type I error probability. The ARL is the average number of samples to be checked in order to obtain a point outside the control limit, which in this paper would be a point above the UCL. For the GPD model, the ARL is given by:

$$\text{ARL} = \frac{1}{\text{Pr(point out of control limit)}} = \frac{1}{\sum_{k=\text{UCL}}^{\infty} \frac{\theta(\theta + \lambda k)^{k-1} e^{-\theta - \lambda k}}{k!}}.$$

(2.23)

He et al. (2006) calculated the ARL values of GPD control chart for various choices of parameters and probability of type I error. Some numerical values of ARL for different α_L, θ, and λ are presented in Table 2.9 taken from He et al. (2006).

For example, with the false alarm probability $\alpha_L = 0.0005$, the ARL is calculated to be 2071 when parameters θ and λ are 0.80 and 0.01, respectively. That is, on average, 2071 samples need to be checked to obtain a point plotted outside the control limit.

A Geometric Chart to Monitor Parameter θ

For the GPD model, we have that $p = P$ (one or more defects in a product) $= 1 - e^{-\theta}$. Monitoring p is equivalent to monitoring the parameter. Hence, the procedure for monitoring the parameter θ is to count the number of products examined until a nonconforming product is detected. This procedure is repeated whenever a nonconforming product is produced. A geometric control chart will then be used to

Table 2.9 Some numerical values of ARL for $\theta_0 = 0.01$, $\lambda_0 = 0.8$, and $\alpha_L = 0.0005$.

θ	$\lambda = 0.50$	$\lambda = 0.75$	$\lambda = 0.80$	$\lambda = 0.85$	$\lambda = 0.90$	$\lambda = 1.00$	$\lambda = 1.50$
0.005	312,409	6,626	4,147	2,775	1,966	1,137	342
0.01	155,491	3,308	2,071	1,386	983	568	171
0.03	50,891	1,097	688	461	327	190	57
0.05	29,984	654	411	276	196	114	35
0.10	14,329	322	203	137	98	57	18
0.15	9,134	212	134	91	65	38	12
0.20	6,552	157	100	68	48	29	9

monitor the count of conforming units. Thus, by using the geometric chart, we could tell whether there is any change in the parameter θ. The cumulative count of conforming items until a nonconforming one is denoted by CCC. Note that CCC should include the last nonconforming item since in practical applications every round of inspection will stop only when a nonconforming item is encountered. Obviously, CCC follows a geometric distribution. The probability of false alarm is defined by α_{ccc}. The control limits for geometric chart are computed by:

$$\text{UCL}_{ccc} = \frac{\ln(\alpha_{ccc}/2)}{\ln(1-p)}$$

$$\text{CL}_{ccc} = \frac{\ln(0.5)}{\ln(1-p)}$$

$$\text{LCL}_{ccc} = \frac{\ln(1-\alpha_{ccc}/2)}{\ln(1-p)}.$$

They also proposed a combine procedure to monitor both parameters of GPD. For further details one can see He et al. (2006).

2.4.3 The Q- and the T-Charts

Shmueli, Minka, Kadane, Borle, and Boatwright (2005) revived a flexible probability distribution called the COM–Poisson distribution that can broadly model count data that are either over- or under-dispersed. Let X denote the number of nonconformities in product is observed a COM-Poisson random variable. Then, X is a COM–Poisson random variable with p.m.f.

$$P(X = x|\lambda, \nu) = \frac{\lambda^x}{(x!)^\nu z(\lambda, \nu)}, \quad \text{for} \quad \lambda > 0, \nu \geq 0, x = 0, 1, 2, \ldots \tag{2.24}$$

where λ is a location parameter, ν is a dispersion parameter, and $z(\lambda, \nu) = \sum_{s=0}^{\infty} \frac{\lambda^s}{(s!)^\nu}$ is a normalizing constant. This distribution encompasses three well-known distributions as special cases: Poisson ($\nu = 1$), geometric ($\nu = 0$, $\lambda < 1$), and Bernoulli ($\nu \to \infty$ with probability ($\lambda/(1 + \lambda)$)). Note that $\nu > 1$ addresses the case of data under-dispersion, whereas $\nu < 1$ implies over-dispersion. The associated moment generating function of X is $M_{X(t)} = E(e^{tX}) = z(\lambda e^t, \nu)/z(\lambda, \nu)$.

Sellers (2012) used the COM–Poisson distribution in Shewhart attribute control charts and shows that the control chart based on the COM–Poisson distribution is flexible and a generalization of the p-chart, the c-chart, and the u-chart. The k-sigma control limits proposed by Sellers (2012) based on location-shifted COM–Poisson distribution are given in Table 2.10.

Table 2.10 Control limits for total number of events and average number of events based on the COM–Poisson distribution, respectively.

	Total number of counts	Average number of counts
LCL	$n\left(\lambda\dfrac{\partial \log (Z(\lambda, v))}{\partial\lambda} + a\right) - k\sqrt{n\left(\dfrac{\partial E(X)}{\partial \log \lambda}\right)}$	$\left(\lambda\dfrac{\partial \log (Z(\lambda, v))}{\partial\lambda} + a\right) - k\sqrt{\dfrac{1}{n}\left(\dfrac{\partial E(X)}{\partial \log \lambda}\right)}$
CL	$n\left(\lambda\dfrac{\partial \log (Z(\lambda, v))}{\partial\lambda} + a\right)$	$\left(\lambda\dfrac{\partial \log (Z(\lambda, v))}{\partial\lambda} + a\right)$
UCL	$n\left(\lambda\dfrac{\partial \log (Z(\lambda, v))}{\partial\lambda} + a\right) + k\sqrt{n\left(\dfrac{\partial E(X)}{\partial \log \lambda}\right)}$	$\left(\lambda\dfrac{\partial \log (Z(\lambda, v))}{\partial\lambda} + a\right) + k\sqrt{\dfrac{1}{n}\left(\dfrac{\partial E(X)}{\partial \log \lambda}\right)}$

Table 2.11 Control chart multiple k required to achieve $\alpha = 0.0020$ for COM–Poisson Q-chart.

	(λ, ν)								
n	(1.5, 0.5)	(1.5, 1.5)	(1.5, 2.0)	(2.5, 0.5)	(2.5, 1.5)	(2.5, 2.0)	(3.0, 0.5)	(3.0, 1.5)	(3.0, 2.0)
3	3.332	3.302	3.287	3.264	3.119	3.113	3.111	3.109	3.107
5	3.329	3.295	3.256	3.210	3.077	3.076	3.075	3.073	3.072
10	3.323	3.285	3.221	3.184	3.049	3.048	3.047	3.043	3.041
15	3.307	3.264	3.185	3.136	3.036	3.035	3.034	3.030	3.029

However, these limits are not practical and misleading for the case of the asymmetric COM–Poisson distribution. Saghir et al. (2013) extended the work of Sellers and developed the probability control limits of the COM–Poisson charts. They also determined the control charting constants k for various values of n and α. Tables 2.11 and 2.12 give some values of control charting constant used in the construction of generalized and flexible control charts, namely, Q-chart and T-Chart for count data, respectively.

It is obvious that the value of the control chart multiplier k is not equal to 3, particularly for small sample size (see Tables 2.11 and 2.12). Thus, the use of control limits proposed by Sellers (2012) is misleading particularly for small subgroup size which is the most practical situation in statistical quality control. For the calculation of k-sigma limits for COM–Poisson charts, the value of k given in Tables 2.11 and 2.12 should be used instead of using $k = 3$. For the fixed value of n, there is not much difference among the values of k for different values of

Table 2.12 Control chart multiple k required to achieve $\alpha = 0.0020$ for COM–Poisson T-chart.

n	(λ, ν)								
	(1.5, 0.5)	(1.5, 1.5)	(1.5, 2.0)	(2.5, 0.5)	(2.5, 1.5)	(2.5, 2.0)	(3.0, 0.5)	(3.0, 1.5)	(3.0, 2.0)
3	3.323	3.291	3.276	3.253	3.110	3.100	3.099	3.109	3.098
5	3.318	3.285	3.247	3.201	3.065	3.063	3.062	3.073	3.060
10	3.313	3.274	3.212	3.172	3.039	3.037	3.035	3.043	3.034
15	3.298	3.253	3.176	3.125	3.027	3.024	3.021	3.030	3.019

parameters. Therefore, for other choices of parameters lying between the pairs given in Tables 2.11 and 2.12, the value of k can be used as the nearest approximation, and it is better than to use $k = 3$. Furthermore, Tables 2.11 and 2.12 argue that $k = 3$ can be used for large l as well as for large n.

For the given value of α, the acceptable risks of false alarm, the UCL and LCL, can be obtained by the solution of these equations:

$$\left.\begin{array}{l} P(Q < \mathrm{LCL}_a) = \displaystyle\sum_{q=0}^{\mathrm{LCL}_a} f(Q) \leq \dfrac{a}{2} \\[2em] P(Q > \mathrm{UCL}_a) = \displaystyle\sum_{q=\mathrm{UCL}_a}^{\infty} f(Q) \leq \dfrac{a}{2} \end{array}\right\}, \tag{2.25}$$

where $f(Q)$ is the probability distribution function of sum of "n" COM–Poisson random variables, i.e. $Q = \sum_1^n X_i$. From Eq. (2.25), one can see that LCL and UCL are $\left(\frac{a}{2}\right)$th and $\left(1 - \frac{a}{2}\right)$th quantiles of the distribution, respectively, and the true probability limits are asymmetrical. The statistical software R can be used to evaluate the COM–Poisson probability function.

The probability control limits are better to use than the symmetric "three sigma" limits for COM–Poisson charts because the COM–Poisson distribution is quite asymmetric. Using power function as performance measure, they concluded that the true probability limits are more powerful to detect a shift in the parameter and provide unbiased power function while k-sigma limits do not. The three sigma limits result in tighter control limits and produce large probability of false alarm than the probability limits. A simple example of this is shown in Table 2.13, where $\lambda = 2.50$, $\upsilon = 0.50$, $n = 3$, and $\alpha = 0.0020$. The false alarm rate associated with probability limits is close to 0.0020, but the associated probability of 3σ limits is nearly 50% larger than specified.

Table 2.13 False alarm probability example for COM–Poisson chart.

Limits	(LCL, UCL)	Specified probability	Associated probability
three sigma	(0.7329, 11.5442)	0.0020	0.0095
Probability	(2.6000, 12.2000)	0.0020	0.0018

The OC Curve

The OC function for the total number of counts (Q-) chart can be defined by the following equation:

$$\beta = \text{Pr(incorrectly accepting the hypothesis of statistical control)}$$
$$\beta = \text{Pr}(\text{LCL} \leq Q < \text{UCL}|\lambda_1)$$
$$\beta = F_Q(\text{UCL}) - F_Q(\text{LCL}) \tag{2.26}$$

where $F_Q(., a)$ is the cumulative density function of COM–Poisson distribution and λ_1 is the shifted value of location parameter when dispersion parameter is fixed at v_0.

Similarly, the OC function for the average number of counts (T-) chart can be defined by the following equation:

$$\beta = \text{Pr(incorrectly accepting the hypothesis of statistical control)}$$
$$\beta = \text{Pr}(\text{LCL} \leq T < \text{UCL}|\lambda_1)$$
$$\beta = F_T(\text{UCL}) - F_T(\text{LCL}) \tag{2.27}$$

where $F_T(., a)$ is the cumulative density function of COM–Poisson distribution and λ_1 is the shifted value of location parameter when dispersion parameter is fixed at v_0.

The power curves for the k-sigma and probability limit schemes for the COM–Poisson T-chart are constructed by Saghir et al. (2013) for various choices of parameter and shift sizes. Figure 2.7, taken from Saghir et al. (2013), presents the power curve of T-chart, for $n = 10$ and $v = 1.5$.

2.5 Other Recent Developments

An attribute control chart, namely, np_{S^2} control chart, to monitor the variability of a process was explored by Ho and Quinino (2013), and the results are described in this paper. A comparison of the proposed np_{S^2} control chart with the S^2 and R

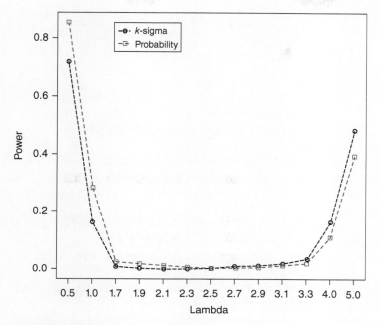

Figure 2.7 Power curve for *T*-chart at *n* = 10 and *v* = 1.5.

control charts was studied. Shafqat, Hussain, Al-Nasser, and Aslam (2018) studied *np*-chart under truncated life test using Burr X & XII, inverse Gaussian (IG), and exponential lifetime-truncated distributions. The performance of *np*-truncated chart was evaluated by ARL, which compares the performance of all distributions. Aslam et al. (2018) designed an attribute control chart for two-stage process control. The in-control ARL was derived, and the out-of-control ARLs were also analyzed according to the process shifts in the first- and/or the second-stage process. Erginel, Sentürk, and Yıldız (2018) used fuzzy attribute control charts, where data are classified into conforming/nonconforming product units to monitor fuzzy fractions of nonconforming units for variable sample sizes and the fuzzy number of nonconforming units for constant sample sizes. Data defined as quality characteristics can be imprecise due to the subjective decisions of the quality control operator. Type-2 fuzzy set theory deals with ambiguity associated with the uncertainty of membership functions by incorporating footprints and modeling high-level uncertainty. In this paper, the structure of an interval type-2 fuzzy *p*-control chart and interval type-2 fuzzy *np*-control chart with constant sample size is developed and applied to real data. The main advantage in using interval type-2 fuzzy sets in control charts is the flexibility allowed in determining control limits for process monitoring by incorporating fuzzy set theory. Therefore, fuzzy control charts with interval type-2 fuzzy numbers afford the decision maker the opportunity to see and detect process defects.

References

Acosta-Mejia, C. A. (1999). Improved *p* charts to monitor process quality. *IIE Transactions, 31*(6), 509–516.

Aebtarm, S., & Bouguila, N. (2001). An empirical evaluation of attribute control charts for monitoring defects. *Expert System with Applications, 38*, 7869–7880.

Albers, W. (2009). Control charts for health care monitoring under over-dispersion. *Metrika, 74*(1), 67–83.

Aslam, M., Azam, M., Kim, K.-J., & Jun, C.-H. (2018). Designing of an attribute control chart for two-stage process. *Measurement and Control, 51*, 285–292.

Braun, W. J. (1999). Run length distribution for estimated attributes charts. *Metrika, 50*, 121–129.

Chakraborti, S., & Human, S. W. (2008). Properties and performance of the c chart for attributes data. *Journal of Applied Statistics, 35*(1), 89–100.

Chan, L. Y., Lai, C. D., Xie, M., & Goh, T. N. (2003). A two-stage decision procedure for monitoring processes with low fraction nonconforming. *European Journal of Operational Research, 150*(2), 420–436.

Chen, N., Zhou, S., Chang, T. S., & Huang, H. (2008). Attribute control charts using generalized zero-inflated Poisson distribution. *Quality and Reliability Engineering International, 24*(7), 793–806.

Duran, R. I., & Albin, S. L. (2009). Monitoring a fraction with easy and reliable settings of the false alarm rate. *Quality and Reliability Engineering International, 25*(8), 1029–1043.

Erginel, N., Şentürk, S., & Yıldız, G. (2018). Modeling attribute control charts by interval type-2 fuzzy sets. *Soft Computing, 22*, 5033–5041.

Famoye, F. (2007). Statistical control charts for shifted generalized Poisson distribution. *Statistical Methods and Applications, 3*(3), 339–354.

From, S. G. (2004). Approximating the distribution of a renewal process using generalized Poisson distributions. *Journal of Statistical Computation and Simulation, 74*, 667–681.

Goh, T. N. (1987). A charting technique for control of low-defective production. *International Journal of Quality and Reliability Management, 4*(1), 53–62.

Goh, T. N., & Xie, M. (2003). Statistical control of a six sigma process. *Quality Engineering, 15*(4), 587–592.

He, B., Xie, M., Goh, T. N., & Tsui, K. L. (2006). On control charts based on the generalized Poisson model. *Quality Technology and Quantitative Management, 3*(4), 383–400.

Ho, L. L., & Quinino, R. C. (2013). An attribute control chart for monitoring the variability of a process. *International Journal of Production Economics, 145*, 263–267.

Jahromi, S. M. Z., Saghaei, A., & Ariffin, M. K. A. (2012). A review on fuzzy control charts for monitoring attribute data. *Applied Mechanics and Materials, 159*, 23–28.

Kaminsky, F. C., Benneyan, J. C., Davis, R. D., & Burke, R. J. (1992). Statistical control charts based on a geometric distribution. *Journal of Quality Technology, 24*(2), 63–69.

Laney, D. B. (2002). Improved control charts for attributes. *Quality Engineering, 14*(2), 531–537.

Lucas, J. M., Davis, D. J., & Saniga, E. M. (2010). Detecting improvement using Shewhart attribute control charts when the lower control limit is zero. *IIE Transactions, 38*(8), 699–709.

Luceno, A. (2005). Recursive characterization of a large family of discrete probability distributions showing extra-Poisson variation. *Statistics, 39*, 261–267.

Mohammed, M. A., Panesar, J. S., Laney, D. B., & Wilson, R. (2003). Statistical process control charts for attribute data involving very large sample sizes: A review of problems and solutions. *BMJ Quality and Safety, 12*, 362–368.

Montgomery, D. C. (2013). *Introduction to statistical quality control* (7th ed.). New York: Wiley.

Nelson, L. S. (1997). Supplementary runs tests for np control charts. *Journal of Quality Technology, 29*(2), 225–227.

Ryan, T. P., & Schwertman, N. C. (1997). Optimal limits for attributes control hcharts. *Journal of Quality Technology, 29*, 86–98.

Saghir, A., & Lin, Z. (2014). Control charts for the dispersed count data: An overview. *Quality and Reliability Engineering International, 31*(5), 725–739.

Saghir, A., Lin, Z., Abbasi, S. A., & Ahmad, S. (2013). The use of probability control limits of the COM-Poisson charts and their applications. *Quality and Reliability Engineering International, 29*, 759–770.

Schwertman, N. C., & Ryan, T. P. (1997). Implementing optimal attributes control charts. *Journal of Quality Technology, 29*(1), 99–104.

Schwertman, N. C., & Ryan, T. P. (1999). Using dual *np*-charts to detect changes. *Quality and Reliability Engineering International, 15*(4), 317–320.

Sellers, K. F. (2012). A generalized statistical control chart for over-or under-dispersed data. *Quality and Reliability Engineering International, 28*, 59–65.

Shafqat, A., Hussain, J., Al-Nasser, A. D., & Aslam, M. (2018). Attribute control chart for some popular distributions. *Communications in Statistics – Theory and Methods, 47*(8), 1978–1988.

Shankar, V., Milton, J., & Mannering, F. (1997). Modeling accident frequencies as zero-altered probability processes: An empirical inquiry. *Accident Analysis and Prevention, 29*, 829–837.

Shewhart, W. A. (1926). Quality control charts. *Bell Systems Technical Journal, 5*(4), 593–603.

Shewhart, W. A. (1927). Quality control. *Bell Systems Technical Journal, 6*(4), 722–735.

Shmueli, G., Minka, T. P., Kadane, J. B., Borle, S., & Boatwright, P. (2005). A useful distribution for fitting discrete data: Revival of the Conway–Maxwell–Poisson distribution. *Applied Statistics, 54*, 127–142.

Szarka, J. L., III, & Woodall, W. H. (2011). A review and perspective on surveillance of Bernoulli processes. *Quality and Reliability Engineering International, 27*, 735–752.

Tang, L. C., & Cheong, W. T. (2006). A control scheme for high-yield correlated production under group inspection. *Journal of Quality Technology, 38*(1), 45–55.

Vaughan, T. S. (1993). Variable sampling interval *np* process control chart. *Communications in Statistics – Theory and Methods, 22*(1), 147–167.

Vries, A. D., & Teneau, J. K. (2010). Application of statistical process control charts to monitor changes in animal production systems. *Journal of Animal Sciences, 88*, 2009–2622.

Wang, H. (2009). Comparison of *p* control charts for low defective rate. *Computational Statistics and Data Analysis, 53*(12), 4210–4220.

Woodall, W. H. (1997). Control charts based on attribute data: Bibliography and review. *Journal of Quality Technology, 29*, 172–183.

Woodall, W. H. (2000). Controversies and contradictions in statistical process control. *Journal of Quality Technology, 32*, 341–350.

Woodall, W. H. (2006). The use of control charts in health-care and public-health surveillance. *Journal of Quality Technology, 38*(2), 89–104.

Woodall, W. H., Tsui, K. L., & Tucker, G. R. (1997). A review of statistical and fuzzy quality control charts based on categorical data. *Frontiers in Statistical Quality Control, 5*, 83–89.

Wu, Z., Yeo, S. H., & Spedding, T. A. (2001). A synthetic control chart for detecting fraction nonconforming increases. *Journal of Quality Technology, 33*(1), 104–111.

Xie, M., & Goh, T. N. (1992). Some procedures for decision making in controlling high yield processes. *Quality and Reliability Engineering International, 8*(4), 355–360.

Xie, M., & Goh, T. N. (1993). Improvement detection by control charts for high yield processes. *International Journal of Quality and Reliability Management, 10*(7), 24–31.

Xie, M., Lu, X. S., Goh, T. N., & Chan, L. Y. (1999). A quality monitoring and decision-making scheme for automated production processes. *International Journal of Quality and Reliability Management, 16*(2), 148–157.

Zhang, M., Peng, Y., Schuh, A., Megahed, F. M., & Woodall, W. H. (2013). Geometric charts with estimated control limits. *Quality and Reliability Engineering International, 29*, 209–223.

3

Variable Control Charts

3.1 Introduction

In statistics the quantitative variable is classified as the discrete and the continuous variables. The collection of data in the form of integers or the whole numbers, such as number of defective items produced in a day, number of employees in a production unit, and number of lots submitted for inspection per day, are the examples of the discrete or discontinuous variable (Bowker & Goode, 1952). The control charts used for monitoring of this form of data are known as the attribute control charts (have been discussed in the previous chapter). Shewhart's charts have been developed for identifying special causes of variations. These causes may be defined as the slowly increased rare changes or the large changes in their effect. Any cause that is present all the time in the process even though it is large is not a special cause (Allen, 2006; Pearn & Kotz, 2006; Trietsch, 1998).

The collection of data in the form of any conceivable values such as volume, weight, and dimension within any observed range such as the weight of manufactured item as 120 and 130 g, an infinite number of weights is possible between 120 and 130 g, a manufactured item may be 10–15 cm long, and the failure time of an electric good may be recorded as 20,000–22,000 hours. A continuous variable is one that can assume each and every possible value in a given interval (Amiri, Noghondarian, & Noorossana, 2014). The control limits describe the variation in the product that is expected to lie in a state of statistical control (Xie, Goh, & Kuralmani, 2012). The present chapter is written for the developments of control chart schemes for such variables.

Ever since the introduction of the notion of the control chart by Shewhart (1931), the quality practitioners are continuously endeavoring for devising a robust

---

Preamble: The most commonly used control charts for process monitoring are the variable control charts. In this chapter various types of charts under the measureable characteristics are highlighted with recent developments and modifications.

Introduction to Statistical Process Control, First Edition. Muhammad Aslam, Aamir Saghir, and Liaquat Ahmad.
© 2021 John Wiley & Sons, Inc. Published 2021 by John Wiley & Sons, Inc.

scheme for the monitoring of the process and the product (Hawkins & Olwell, 2012). Control charts are used to ensure that the manufactured product meet the requirements, set by the consumers, and are improved on a continuous scale according to consumer's desires. Since the variability in the standards or parameters of the products is an important source of poor quality of the products. In addition, if the variability is higher than the mean, then the mean chart is meaningless in any process. Therefore, the dispersion and the location must be considered side by side. So, the purposeful monitoring of the targeted variable is monitored for the possible improvement or decline through control chart technique (Mason & Young, 2002; Oakland, 2008). The mean monitoring of the interested quality characteristic is a common practice, but the monitoring of the variability in addition to mean is necessary to monitor the process efficiently (Montgomery, 2009). The important and commonly adopted control charts by quality control personnel are described in the subsequent sections.

3.2 $\bar{X}$ Control Charts

Control charts help us in monitoring the behavior of the production process to determine whether it is stable. A stable process is one that remains consistent with the center line having minimum variation. The center line of the process is calculated which is the mean of $\bar{X}$ and is denoted by $\bar{\bar{X}}$ (Rahim & Banerjee, 1993). Shewhart introduced this simple measure that is the best way to estimate the center line around which the observations of the running process are posted. With this object standard quality control charts have been being used by the quality control practitioners for decades. By fixing the center line as the target line of the control chart, the more reliable deviations from the target can easily be detected. We normally expect from every process that it should not only be in control but also be centered. Occasionally, we fail to adjust the process at the required target, but we know to adjust it for any particular value. The monitoring of process through control chart is of practical use even if we do not know its distribution. In such situations the center line is drawn close to the target line. If the distribution of the observations is ignorant, then we are unable to estimate the false alarm rate (FAR). Shewhart demonstrated that use of subgroup of size equal to 4 or 5 provides the normally distributed data even if the observations are highly skewed. Charts developed using the dispersion of the observations are more sensitive to deviation from normality (Castagliola, Celano, & Psarakis, 2011). For a set of samples of optimal size observed from a normal population with mean μ and standard deviation σ, the traditional mean Shewhart's control chart commonly known as $\bar{X}$ chart, the upper control limit (UCL), and the lower control limit (LCL) are constructed by mean $\pm 3\hat{\sigma}$, which is the estimated value of the population standard deviation.

The valid and objective control limit estimation is directly related with the estimation of the population standard deviation. Clearly, the main disadvantage of this scheme is the unavailability of the population standard deviation σ. A huge amount of literature is available for different nice estimations of the population standard deviation. Three different estimations of the population standard deviation have been proposed by Chakraborti, Human, and Graham (2008). These three unbiased estimations are based on (i) the average of sample ranges, (ii) the average of sample variances, and (iii) the average of sample standard deviations.

3.2.1 Construction of $\overline{X}$ and R Charts

Let n be the subgroup size and m be the total number of subgroups such that the variation within subgroups is smaller than between groups. Further let X_{ij} be the interested quality characteristic of the ith unit and the jth subgroup, where $i = 1, 2, 3, ..., n$ and $j = 1, 2, 3, ..., m$.

King (1954) was the first who developed the $\overline{X}$ control chart by adopting the grand average

$$\overline{\overline{X}} = \sum_{i=1}^{m} \overline{X}_i / m \tag{3.1}$$

$$\overline{\overline{X}} = \sum_{j=1}^{m} \sum_{i=1}^{n} X_{ij} / mn \tag{3.2}$$

and the average of the ranges of the subsamples $\overline{R} = \sum_{i=1}^{m} R_i / m$ to estimate the process mean and standard deviation, respectively. The three-sigma control limits using these estimates can be developed as

$$\text{UCL} = \overline{\overline{X}} + A_2 \overline{R} \tag{3.3}$$
$$\text{LCL} = \overline{\overline{X}} - A_2 \overline{R} \tag{3.4}$$

where A_2 is $3/d_2\sqrt{n}$ and a function that depends only on n, the subgroup size.

King (1954) proposed the control chart constant as $C = K_m/(d_2\sqrt{n})$ with the control limits as

$$\text{UCL} = \overline{\overline{X}} + C\overline{R} \tag{3.5}$$
$$\text{LCL} = \overline{\overline{X}} - C\overline{R} \tag{3.6}$$

With the claim that these limits depend not only on the subsample size but also on the total number of subsamples, King (1954) developed the approximate values of the chart constant C using simulations for $m = 3(1)25$ and $n = 2, 3, 4, 5$, and 10 with the false alarm probability equal to 0.05. King (1954) perceived that as m increases, the constant C approaches to A_2. The above mentioned ideas have been illustrated using the following example.

Example 3.1 The net weight (in ounce) of a dry bleach product to be monitored by the $\overline{X}$ and R control charts using a sample of size $n = 5$. Data for 20 preliminary samples are given in Table 3.1.

The three-sigma control limits for $\overline{X}$ chart are

$$\text{UCL} = \overline{\overline{X}} + A_2\overline{R} \qquad (3.7)$$

and

$$\text{LCL} = \overline{\overline{X}} - A_2\overline{R} \qquad (3.8)$$

Table 3.1 Net weight (in ounce) of a dry bleach product.

Sample no.	x_1	x_2	x_3	x_4	x_5	Average	Range	S
1	15.8	16.3	16.2	16.1	16.6	16.20	0.8	0.2915
2	16.3	15.9	15.9	16.2	16.4	16.14	0.5	0.2302
3	16.1	16.2	16.5	16.4	16.3	16.30	0.4	0.1581
4	16.3	16.2	16.9	16.4	16.2	16.20	0.5	0.1871
5	16.1	16.1	16.4	16.5	16.0	16.22	0.5	0.2168
6	16.1	15.8	16.7	16.6	16.4	16.32	0.9	0.3701
7	16.1	16.3	16.5	16.1	16.5	16.30	0.4	0.2000
8	16.2	16.1	16.2	16.1	16.3	16.18	0.2	0.0837
9	16.3	16.2	16.4	16.3	16.5	16.34	0.3	0.1140
10	16.6	16.3	16.4	16.1	16.5	16.38	0.5	0.1924
11	16.2	16.4	15.9	16.3	16.4	16.24	0.5	0.2074
12	15.9	16.6	16.7	16.2	16.5	16.38	0.8	0.3271
13	16.4	16.1	16.6	16.4	16.1	16.32	0.5	0.2168
14	16.5	16.3	16.2	16.3	16.4	16.34	0.3	0.1140
15	16.4	16.1	16.3	16.2	16.2	16.24	0.3	0.1140
16	16.0	16.2	16.3	16.3	16.2	16.20	0.3	0.1225
17	16.4	16.2	16.4	16.3	16.2	16.30	0.2	0.1000
18	16.0	16.2	16.4	16.5	16.1	16.24	0.5	0.2074
19	16.4	16.0	16.3	16.4	16.4	16.30	0.4	0.1732
20	16.4	16.4	16.5	16.0	15.8	16.22	0.7	0.3033
Means						16.2680	0.4750	0.1965

Source: Data from Montgomery (2009, p. 273).

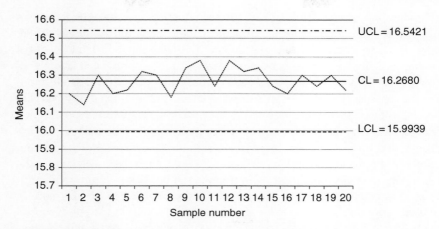

Figure 3.1 $\overline{X}$ chart of net weight (in ounce) of a dry bleach product.

$\overline{\overline{X}} = 16.2680$, $\overline{R} = 0.4750$, and $A_2 = 0.5770$ Then UCL = 16.5421 and LCL = 15.9939 (Figure 3.1).

The control limits of the R chart can be defined as

$$\text{UCL} = D_4\overline{R} \tag{3.9}$$

$$\text{Central line} = \overline{R}$$

$$\text{LCL} = D_3\overline{R} \tag{3.10}$$

where $D_3 = 0$ and $D_4 = 2.114$ are the R-chart constants (Figure 3.2).
Thus

$$\text{UCL} = 2.114*0.475$$

$$\text{UCL} = 1.0042$$

$$\text{Central line} = 0.4750$$

$$\text{LCL} = 0$$

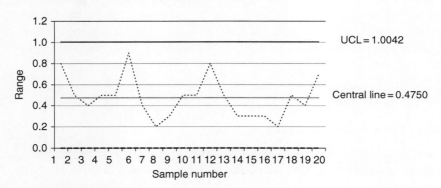

Figure 3.2 R chart for the net weight (in ounce) of a dry bleach product.

The development of control chart for a production process consists of two phases: the phase I and phase II. In phase I the control limits are developed, which are also known as the trial control limits, from the m initial set of samples or subsamples of reasonable size of 15–25 with each sample of rational size of 3–5. Normally, the control limits of phase I are set to monitor the process for future when the control limits thus developed are used for some current behavior of the process. For this purpose we develop the $\overline{X}$ and R charts and plot the current observation on the chart and examine the behavior of the process by resulting display. Thus if all the observations fall within the control limits and no systematic pattern is seen, then we conclude that the process was in control and these control limits can be used in future to monitor the process (Chandara, 2001). There may be situations when one or more of the observations do not show the in-control process or falls in the out-of-control region then our hypothesis of in-control process will be rejected and we have to develop the revised trial control limits. In such situation the out-of-control points are critically observed for the unusual changes and thus such points are discarded and the control limits are calculated with the remaining observations (Montgomery, 2009). This scenario may portray the tightened control limits, and this process is continued till all the observations fall within the UCL and LCL.

Periodic revision of control charts plays an important role in the efficient monitoring of the process. By the law of nature, no production process remains at the same level for a long time. So it becomes mandatory to revise the control limits each day, each week, each month, or after 100–200 units. For revision of the control limits, it must be kept in mind that at least 20–25 subsamples or subgroups be used for the development of the periodic revision of the control limits. If the R chart shows an in-control behavior of the production process, then the central line (CL) of the $\overline{X}$ chart can be altered or adjusted to new levels that can also be set as the target value, $\overline{\overline{X}}_0$ (Liu, Chou, & Chen, 2002). When the R chart exhibits the out-of-control process, then the process is analyzed for the out-of-control points and such points are removed from the calculations and new limits are developed for the remaining observations resulting in tightening the control limits. In such situations, the $\overline{X}$ limits are also revised and readjusted.

3.2.2 Phase II Control Limits

When the reliable control limits are achieved from phase I, the future monitoring of the process is based on these limits, which are known as the phase-II monitoring control limits. The online monitoring process is then used for the rapid detection of the unusual changes in the process by using the following sensitizing rules or the Western Electric rules.

Shewhart's control chart sensitizing rules are used to declare the production process whether it is out of control. These rules are as follows (Montgomery, 2009):

1) Any observation falling outside the UCL or LCL (three-sigma limit).
2) Any two out of three consecutive observations fall on the same side of the CL at more than 2 sigma but still in the in-control limit.
3) Any four out of five consecutive observations fall on the same side of the CL at more than 1 sigma.
4) Eight consecutive observations fall on the same side of the CL.
5) Six points steadily increasing or decreasing in a row.
6) Fifteen points in a row within ±1 sigma.
7) Fourteen points alternating up and down in a row.
8) Eight points on both side of the center line with 1 sigma in a row.
9) A nonrandom or unusual pattern in the data.
10) One or more points in the vicinity of the warning or the control limits.

3.2.3 Construction of $\overline{X}$ Chart for Burr Distribution Under the Repetitive Sampling Scheme

The monitoring of mean value can be administered from any phenomena (Kruger & Xie, 2012; Schilling and Neubauer, 2009). For example, the mean or average monitoring of the process can be developed for any normal or non-normal distribution for any sampling scheme. Let us develop a control chart technique for the Burr type XII for non-normal distribution under the repetitive sampling scheme (Yan, Liu, & Azam, 2017). The detail of the repetitive sampling scheme has been explained in detail in Chapter 6. Two steps can be adopted for the construction of the $\overline{X}$ or mean chart that can be elaborated as

Step I: Select a random sample of size n from the manufacturing process. Calculate mean $\overline{X}$.

Step II: Declare the process as out of control if $\overline{X} \geq \text{UCL}_1$ or $\overline{X} \leq \text{LCL}_1$. The two limits UCL_1 and LCL_1 are also known as the outer control limits. If $\overline{X} \leq \text{UCL}_2$ or $\overline{X} \geq \text{LCL}_2$, then declare the process as in control. Here the two limits UCL_2 and LCL_2 are also known as the inner control limits. If the value of the variable $\overline{X}$ lies between the upper outer and upper inner class limits $(\text{UCL}_2 \leq \overline{X} \leq \text{UCL}_1)$ or the variable $\overline{X}$ lies between the lower outer and lower inner class limits $(\text{LCL}_2 \leq \overline{X} \leq \text{LCL}_1)$, then the decision is pending for repeating the sample (resampling is necessary to reach any decision) (Maedh, Ahmad, & Khan,

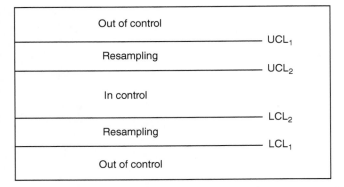

Figure 3.3 Repetitive group sampling scheme procedure (Ahmad, Aslam, & Jun, 2014a).

2017). In this case we go to Step I and repeat the process. A typical repetitive group sampling scheme chart has been given in Figure 3.3.

The cumulative distribution function of the Burr distribution may be defined as

$$f(y) = 1 - \frac{1}{(1 + y^c)^q} \quad y \geq 0$$

where c and q represent the skewness and kurtosis of the Burr distribution, respectively.

The four control limits of the repetitive group sampling scheme for the Burr distribution can be constructed as

$$UCL_1 = \mu_0 + K_1 \frac{\sigma}{\sqrt{n}} \tag{3.11}$$

$$LCL_1 = \mu_0 - K_1 \frac{\sigma}{\sqrt{n}} \tag{3.12}$$

$$UCL_2 = \mu_0 + K_2 \frac{\sigma}{\sqrt{n}} \tag{3.13}$$

$$LCL_2 = \mu_0 - K_2 \frac{\sigma}{\sqrt{n}} \tag{3.14}$$

where μ_0 is the overall process mean, K_1 and K_2 are the control chart coefficients, and σ is the process standard deviation. The efficiency in terms of early and quick detection of out-of-control process can be evaluated by the average run length (ARL) values of the proposed technique. We can compute the values of ARL for the Burr distribution using the following steps:

Let $\overline{X}$ be the sample mean and S the standard deviation, then we can develop the following relation of $\overline{X}$ as

$$\overline{X} = \mu_0 + (Y - M)\frac{\sigma}{S\sqrt{n}} \tag{3.15}$$

where Y is the random variable of the Burr distribution and M is the mean of the Burr distribution. As we know that the ARL of the in-control process (ARL_0) is defined as

$$\text{ARL}_0 = 1 - \frac{1}{P_{\text{in}}(P)} \tag{3.16}$$

where $P_{\text{in}}(P)$ is the probability that the process is declared as in control for the Burr distribution and is calculated using the relation

$$P_{\text{in}}(P) = \frac{\left(\frac{1}{[1+(M-K_2 S)^c]^q} - \frac{1}{[1+(M+K_2 S)^c]^q}\right)}{1 - \left[\left(\frac{1}{[1+(M-K_1 S)^c]^q} - \frac{1}{[1-(M-K_2 S)^c]^q}\right) + \left(\frac{1}{[1+(M+K_2 S)^c]^q} - \frac{1}{[1+(M+K_1 S)^c]^q}\right)\right]} \tag{3.17}$$

And the ARL of out-of-control process (ARL_1) is defined as

$$\text{ARL}_1 = 1 - \frac{1}{P_{\text{in}}^*} \tag{3.18}$$

where P_{in}^* is the probability that the process is declared as out of control for the Burr distribution and is calculated using the relation

$$P_{\text{in}}^* = \frac{\left(\frac{1}{[1+(M-(K_2 S+CS\sqrt{n}))^c]^q} - \frac{1}{[1+(M+(K_2 S-CS\sqrt{n}))^c]^q}\right)}{1 - \left[\left(\frac{1}{[1+(M-(K_1 S+CS\sqrt{n}))^c]^q} - \frac{1}{[1+(M-(K_2 S+CS\sqrt{n}))^c]^q}\right) + \left(\frac{1}{[1+(M+(K_2 S-CS\sqrt{n}))^c]^q} - \frac{1}{[1+(M+(K_1 S-CS\sqrt{n}))^c]^q}\right)\right]} \tag{3.19}$$

The average sample size (ASS) of the process for the Burr distribution can be computed as

$$\text{ASS}(P) = \frac{n}{1 - \left[P\left(\frac{1}{[1+(M-(K_1 S\sqrt{n}))^c]^q} - \frac{1}{[1+(M-(K_2 S\sqrt{n}))^c]^q}\right) + P\left(\frac{1}{[1+(M+(K_2 S\sqrt{n}))^c]^q} - \frac{1}{[1+(M+(K_1 S\sqrt{n}))^c]^q}\right)\right]} \tag{3.20}$$

ARLs of $\overline{X}$ control chart for Burr distribution when $n = 10, 20$ and $r_0 = 100, 200,$ and 300 (Azam, Ahmad, & Aslam, 2016) (Table 3.2).

Control charts are constructed to monitor the variations in the manufacturing items (Kang, Lee, Seong, & Hawkins, 2007; Tong & Chen, 1991). There are two types of variations in any manufacturing/production process. These variations are the common cause variations and the special cause variations (Lu & Reynolds, 1999). The common cause variations are natural beyond the control of human

Table 3.2 Construction of $\overline{X}$ chart for capability index under the repetitive sampling scheme.

Shift (f)	ARL$_0$ = 100, k_1 = 2.8132, k_2 = 0.6271 ASS$_0$ = 21.12		ARL$_0$ = 200, k_1 = 3.9151, k_2 = 0.2997 ASS$_0$ = 42.29		ARL$_0$ = 300, k_1 = 3.9524, k_2 = 0.5480 ASS$_0$ = 23.97	
	ASS$_1$	ARL$_1$	ASS$_1$	ARL$_1$	ASS$_1$	ARL$_1$
0.0	21.12	100.03	42.29	200.04	23.97	300.03
0.1	21.52	51.23	43.10	43.55	24.52	67.45
0.2	23.50	23.25	45.56	13.11	25.51	20.59
0.3	27.55	10.34	50.55	4.99	30.35	7.55
0.4	32.45	4.59	51.30	2.40	33.64	3.44
0.5	35.45	2.35	45.24	1.50	33.55	1.91
1.0	11.11	1.00	11.00	1.00	10.95	1.00
1.5	5.57	1.00	5.50	1.00	5.99	1.00
2.0	5.52	1.00	5.55	1.00	5.29	1.00
2.5	5.51	1.00	5.27	1.00	5.52	1.00
3.0	5.59	1.00	5.43	1.00	5.42	1.00

Shift (f)	ARL$_0$ = 100, k_1 = 2.9731, k_2 = 0.3915 ASS$_0$ = 65.11		ARL$_0$ = 200, k_1 = 3.0658, k_2 = 06479 ASS$_0$ = 41.27		ARL$_0$ = 300, k_1 = 3.9686, k_2 = 0.5954 ASS$_0$ = 44.52	
	ASS$_1$	ARL$_1$	ASS$_1$	ARL$_1$	ASS$_1$	ARL$_1$
0.0	55.11	100.00	41.27	200.00	44.52	300.07
0.1	59.05	34.25	43.55	55.45	45.59	41.04
0.2	53.94	10.34	53.90	15.94	55.11	9.34
0.3	101.45	3.23	70.12	5.19	63.76	3.04
0.4	92.51	1.47	75.42	1.90	60.37	1.51
0.5	60.10	1.09	57.46	1.15	45.27	1.12
1.0	11.75	1.00	11.59	1.00	12.57	1.00
1.5	11.79	1.00	13.15	1.00	12.55	1.00
2.0	11.71	1.00	11.55	1.00	10.46	1.00
2.5	15.91	1.00	15.72	1.00	15.77	1.00
3.0	19.99	1.00	19.99	1.00	19.92	1.00

Table 3.2 (Continued)

Average run lengths of proposed control charts when $n = 30$, 40 and $r_0 = 100$, 200, 300

Shift (f)	ARL$_0$ = 100, k_1 = 2.7638, k_2 = 0.7374 ASS$_0$ = 55.17		ARL$_0$ = 200, k_1 = 2.6920, k_2 = 0.8970 ASS$_0$ = 47.42		ARL$_0$ = 300, k_1 = 3.1334, k_2 = 0.8662 ASS$_0$ = 48.79	
	ASS$_1$	ARL$_1$	ASS$_1$	ARL$_1$	ASS$_1$	ARL$_1$
0.0	55.17	100.00	47.42	200.00	45.79	300.02
0.1	59.92	29.90	51.55	53.55	53.34	74.90
0.2	77.01	7.55	67.10	12.22	70.45	14.65
0.3	93.35	2.29	57.45	3.04	93.75	3.21
0.4	75.04	1.22	75.00	1.34	52.95	1.34
0.5	47.20	1.03	45.57	1.05	49.55	1.05
1.0	15.55	1.00	19.21	1.00	15.94	1.00
1.5	17.55	1.00	17.75	1.00	17.41	1.00
2.0	25.33	1.00	27.57	1.00	25.97	1.00
2.5	30.00	1.00	29.99	1.00	29.99	1.00
3.0	30.00	1.00	30.00	1.00	30.00	1.00

Shift (f)	ARL$_0$ = 100, k_1 = 2.7388, k_2 = 0.8043 ASS$_0$ = 65.51		ARL$_0$ = 200, k_1 = 2.9679, k_2 = 0.8790 ASS$_0$ = 64.22		ARL$_0$ = 300, k_1 = 3.9929, k_2 = 0.6744 ASS$_0$ = 79.86	
	ASS$_1$	ARL$_1$	ASS$_1$	ARL$_1$	ASS$_1$	ARL$_1$
0.0	55.51	100.00	64.22	200.01	79.55	300.04
0.1	75.42	24.75	72.10	42.75	55.92	22.27
0.2	102.91	5.25	100.59	7.51	112.97	3.53
0.3	112.55	1.53	119.50	1.57	107.54	1.41
0.4	74.45	1.05	79.24	1.10	69.57	1.05
0.5	44.17	1.00	44.59	1.01	43.59	1.00
1.0	25.05	1.00	25.52	1.00	25.54	0.75
1.5	25.27	1.00	25.53	1.00	21.52	0.99
2.0	39.95	1.00	39.91	1.00	39.07	1.00
2.5	40.00	1.00	40.00	1.00	40.00	1.00
3.0	55.51	1.00	40.00	1.00	40.00	1.00

Source: Azam et al. (2016).

beings, whereas the latter type of variation is of prime importance in the quality control literature (Chen & Yeh, 2011; Koo & Case, 1990). There are so many reasons behind the special cause of variations including raw material, machine adjustment, machine operator, and temperature. The notion of control chart was introduced by Shewhart in 1920s to address this type of variation in the production processes. The process capability index or process capability ratio, is a term used for the process improvement, is the study of the ability of a process to produce the items within the specification limits. In this section we describe a procedure for constructing $\overline{X}$ chart for the process capability index C_p under the repetitive group sampling scheme. A control chart developed by Subramani and Balamurali (2012) for the $\overline{X}$ chart is used for the repetitive group sampling scheme for efficient monitoring of the special cause of variations. This type of control chart may be written as

$$\mathrm{LCL}_{\overline{X}} = \overline{\overline{X}} - A_2^* \frac{T}{C_p} \tag{3.21}$$

$$\mathrm{UCL}_{\overline{X}} = \overline{\overline{X}} + A_2^* \frac{T}{C_p} \tag{3.22}$$

where $\overline{\overline{X}}$ is overall process mean, $A_2^* = A_2 \frac{d_2}{6}$, $T = \mathrm{USL} - \mathrm{LSL}$, and $C_p = \frac{\mathrm{USL} - \mathrm{LSL}}{6\sigma}$.

The abovementioned limits were developed by Subramani and Balamurali (2012) for the single sampling scheme. Now we develop the modified control chart for the repetitive sampling scheme. The operational procedure of the repetitive sampling scheme can be stated in the following steps:

Step I: We select a random sample of size n from the production process and calculate $\overline{X} = \sum_{i=1}^{n} X_i/n$ the sample mean and $\overline{R} = \sum_{i=1}^{m} R_i/m$ the mean of the sample range and $C_p = \frac{\mathrm{USL} - \mathrm{LSL}}{6\frac{\overline{R}}{d_2}}$.

Step II: The process is declared as out of control if $\overline{X} \geq \mathrm{UCL}_1$ or $\overline{X} \leq \mathrm{LCL}_1$, and the process is declared as in control if $\mathrm{LCL}_2 \leq \overline{X} \leq \mathrm{UCL}_2$. If this is not the case, then go to Step I.

In repetitive sampling we have to define four control limits, two above the center line and two above the center line, also known as the inner and outer control limits (Ahmad, Aslam, & Jun, 2016). The four control limits with two coefficients A_{21}^* and A_{22}^* can be described as

$$\mathrm{LCL}_1 = \overline{\overline{X}} - A_{21}^* \frac{T}{C_p} \tag{3.23}$$

$$\mathrm{UCL}_1 = \overline{\overline{X}} + A_{21}^* \frac{T}{C_p} \tag{3.24}$$

$$\text{LCL}_2 = \overline{\overline{X}} - A_{22}^* \frac{T}{C_p} \tag{3.25}$$

$$\text{LCL}_2 = \overline{\overline{X}} + A_{22}^* \frac{T}{C_p} \tag{3.26}$$

where $A_{21}^* = A_{21} \frac{d_2}{6}$ and $A_{22}^* = A_{22} \frac{d_2}{6}$.

The ARL of the in-control process denoted by ARL_0, which is the mean of the run length distribution, can be defined as

$$\text{ARL}_0 = \frac{1}{\alpha} \tag{3.27}$$

where α is the probability of type I error.

The ARL of the out-of-control process denoted by ARL_1 can be defined as

$$\text{ARL}_1 = \frac{1}{1 - \beta} \tag{3.28}$$

where β is the probability of type II error.

For the repetitive sampling control chart, the in-control probability denoted by P_{ins} can be written as

$$P_{\text{ins}} = P\left(\text{LCL}_2 \le \overline{X} \le \text{UCL}_2\right) \tag{3.29}$$

The probability of out-of-control process denoted by P_{out} can be written as

$$P_{\text{out}} = P\left(\overline{X} \ge \text{UCL}_1\right) + P\left(\overline{X} \le \text{LCL}_1\right) \tag{3.30}$$

The probability of repetition denoted by P_{rep} can be written as

$$P_{\text{rep}} = P\left(\text{UCL}_2 \le \overline{X} \le \text{UCL}_1\right) + P\left(\text{LCL}_1 \le \overline{X} \le \text{LCL}_2\right) \tag{3.31}$$

Thus the ASS and ARL_0 can be computed as

$$\text{ASS} = \frac{n}{1 - \{\Phi(A_{21}d_2\sqrt{n}) - \Phi(A_{22}d_2\sqrt{n}) + \Phi(-A_{22}d_2\sqrt{n}) - \Phi(-A_{21}d_2\sqrt{n})\}} \tag{3.32}$$

$$\text{ARL}_0 = \frac{1}{1 - \dfrac{\Phi(A_{22}d_2\sqrt{n}) - \Phi(-A_{22}d_2\sqrt{n})}{[\{\Phi(A_{22}d_2\sqrt{n}) - \Phi(-A_{22}d_2\sqrt{n})\} + \{1 - \Phi(A_{21}d_2\sqrt{n}) + \Phi(-A_{21}d_2\sqrt{n})\}]}} \tag{3.33}$$

The ARL of the shifted process denoted by ARL_1 when a shift of the form $\mu_1 = \mu + k\sigma$ for the single sampling scheme can be defined as

$$P_{\text{ins},\mu_1} = P\left(\text{LCL}_2 \leq \overline{X} \leq \text{UCL}_2 \,|\, \mu + k\sigma\right) \tag{3.34}$$

The probability of declaring the shifted process as out of control for single sampling can be written as

$$P_{\text{outs},\mu_1} = P\left(\overline{X} \geq \text{UCL}_1 \,|\, \mu_1 = \mu + k\sigma\right) + P\left(\overline{X} \leq \text{LCL}_1 \,|\, \mu_1 = \mu + k\sigma\right) \tag{3.35}$$

The probability of repetitive sampling for the shifted process P_{reps,μ_1} can be written as

$$P_{\text{reps},\mu_1} = P\left(\text{UCL}_2 \leq \overline{X} \leq \text{UCL}_1 \,|\, \mu_1 = \mu + k\sigma\right) + P\left(\text{LCL}_1 \leq \overline{X} \leq \text{LCL}_2 \,|\, \mu_1 = \mu + k\sigma\right) \tag{3.36}$$

The ASS of the shifted process is given as (Table 3.3)

$$\text{ASS}_1 = \frac{n}{1 - \{\Phi(k\sqrt{n} + A_{21}d_2\sqrt{n}) - \Phi(k\sqrt{n} + A_{22}d_2\sqrt{n}) + \Phi((k\sqrt{n} - A_{22}d_2\sqrt{n}) - \Phi(k\sqrt{n} - A_{21}d_2\sqrt{n}))\}} \tag{3.37}$$

Table 3.3 ARL$_1$ values using the sample sizes n = 5, 10, and 20 when r_0 = 200, 300, and 375 (Ahmad, Aslam, & Jun, 2014b).

	$n = 5$					
	$A_{21} = 0.5724$ $A_{22} = 0.1550$		$A_{21} = 0.5970$ $A_{22} = 0.1515$		$A_{21} = 0.6300$ $A_{22} = 0.0989$	
Shift (δ)	ARL$_1$	ASS$_1$	ARL$_1$	ASS$_1$	ARL$_1$	ASS$_1$
0	200.27	8.58	300.24	8.75	375.19	12.69
0.12	143.16	8.81	209.64	9.00	252.91	13.10
0.13	135.97	8.85	198.47	9.04	238.28	13.17
0.14	128.84	8.90	187.45	9.09	223.94	13.25
0.15	121.84	8.95	176.67	9.14	210.00	13.33
0.20	90.00	9.24	128.32	9.45	148.64	13.85
0.30	46.32	10.11	63.89	10.38	70.21	15.41
0.40	23.48	11.39	31.33	11.77	32.62	17.79
0.50	12.11	13.09	15.59	13.67	15.36	21.08
0.60	6.50	15.09	8.02	16.01	7.50	25.15
0.70	3.72	17.00	4.38	18.45	3.94	29.17
0.80	2.35	18.15	2.63	20.20	2.32	31.44
0.90	1.67	17.93	1.79	20.40	1.59	30.42
1	1.33	16.42	1.38	18.87	1.27	26.51

Table 3.3 (Continued)

Shift (δ)	$A_{21} = 0.3100$ $A_{22} = 0.0705$		$A_{21} = 0.3238$ $A_{22} = 0.0670$		$A_{21} = 0.3308$ $A_{22} = 0.0662$	
			$n = 10$			
0	200.01	19.61	300.22	20.52	375.61	25.16
0.12	106.77	20.76	153.59	21.77	127.94	27.20
0.13	97.75	20.96	139.90	21.99	111.97	27.57
0.14	89.28	21.19	127.14	22.24	97.90	27.97
0.15	81.39	21.43	115.31	22.50	85.55	28.41
0.20	50.33	22.91	69.55	24.13	43.68	31.14
0.30	18.88	27.41	24.79	29.20	12.12	39.61
0.40	7.42	33.81	9.19	36.76	3.99	50.36
0.50	3.30	40.13	3.82	45.08	1.82	54.68
0.60	1.82	41.77	1.96	48.55	1.22	46.99
0.70	1.29	36.73	1.33	43.25	1.06	36.13
0.80	1.10	28.99	1.11	33.73	1.01	28.44
0.90	1.04	22.36	1.04	25.39	1.00	24.03
1	1.01	17.72	1.01	19.60	1.00	21.75

Shift (δ)	$A_{21} = 0.1802$ $A_{22} = 0.0423$		$A_{21} = 0.1880$ $A_{22} = 0.0409$		$A_{21} = 0.1927$ $A_{22} = 0.0387$	
			$n = 20$			
0	200.08	38.26	300.45	39.43	375.36	41.38
0.12	67.86	42.79	95.48	44.28	114.58	46.61
0.13	59.23	43.60	82.78	45.16	98.93	47.57
0.14	51.63	44.50	71.68	46.13	85.30	48.62
0.15	44.97	45.47	62.01	47.18	73.49	49.77
0.20	22.51	51.54	29.99	53.85	34.79	57.07
0.30	6.09	68.95	7.48	74.15	8.27	79.90
0.40	2.20	81.87	2.45	92.59	2.56	102.26
0.50	1.28	71.19	1.32	82.84	1.33	92.22
0.60	1.06	50.55	1.07	57.77	1.07	63.18
0.70	1.01	36.00	1.01	39.74	1.01	42.44
0.80	1.00	27.90	1.00	29.81	1.00	31.16
0.90	1.00	23.66	1.00	24.62	1.00	25.31
1	1.00	21.55	1.00	22.01	1.00	22.35

$$ARL_1 = \frac{1}{1 - \frac{\Phi(k\sqrt{n}+A_{22}d_2\sqrt{n})-\Phi(k\sqrt{n}-A_{22}d_2\sqrt{n})}{\left[\left\{\Phi(k\sqrt{n}+A_{22}d_2\sqrt{n})-\Phi(k\sqrt{n}-A_{22}d_2\sqrt{n})\right\}+\left\{1-\Phi(k\sqrt{n}+A_{22}d_2\sqrt{n})+\Phi(k\sqrt{n}-A_{21}d_2\sqrt{n})\right\}\right]}} \tag{3.38}$$

3.3 Range Charts

Monitoring of production processes using only the process average may lead to erroneous conclusions about the process. In addition to process mean monitoring, the process dispersion or spread monitoring is equally important for effective process monitoring (Gan, 1995; Prajapati & Mahapatra, 2009). Thus the range, quartile deviation, variance, and standard deviation control charts based on sample data of the process are commonly developed in the quality control literature (Chang & Gan, 2004; Chao-Wen & Reynolds, 1999; Chen & Cheng, 1998).

Range is simply the difference between the maximum value and the minimum value of the data set; therefore, R chart is simple and easy to understand and construct, so this chart is the most commonly used chart in practice. But the R chart is not much efficient. Its efficiency decreases rapidly as the sample size increases and the quality control personnel discourages its use for a sample of size greater than 5.

3.4 Construction of *S*-Chart

The construction of S-chart or standard deviation chart is based on the assumption of normal approximation of the sampling distribution of sample standard deviation, S (Khoo, 2005). The three-sigma control limits for the S-chart are constructed as (Montgomery, 2009)

$$UCL = \bar{S} + 3\hat{\sigma}_s \tag{3.39}$$

and

$$LCL = \bar{S} - 3\hat{\sigma}_s \tag{3.40}$$

where $\hat{\sigma}_s$ is an estimate of the population standard deviation. It is estimated as

$$\sigma_s = \sigma/\sqrt{2(n-1)} \tag{3.41}$$

As we know that the unbiased estimate of $\sigma = \bar{S}/c(n)\sqrt{2(n-1)}$

Thus the control limits may be constructed as

$$UCL = \overline{S} + 3\overline{S}/c(n)\sqrt{2(n-1)} \tag{3.42}$$

and

$$LCL = \overline{S} - 3\overline{S}/c(n)\sqrt{2(n-1)} \tag{3.43}$$

These control limits may be written as

$$UCL = B_4\overline{S} \tag{3.44}$$

$$LCL = B_3\overline{S} \tag{3.45}$$

where $B_3 = -3\overline{S}/c(n)\sqrt{2(n-1)}$
   and

$$B_4 = 3\overline{S}/c(n)\sqrt{2(n-1)}$$

The values of B_3 and B_4 can be seen from the tables for different values of n. and

Using the observations of Table 3.1 related to net weight (in ounce) of a dry bleach product data, we can construct the UCL and LCL of the S-chart as (Figure 3.4)

$$UCL = B_4\overline{S} \tag{3.46}$$

$$UCL = 2.089*0.1965$$

$$= 0.4104885$$

$$CL = 0.1965$$

$$LCL = B_3\overline{S} \tag{3.47}$$

$$LCL = 0$$

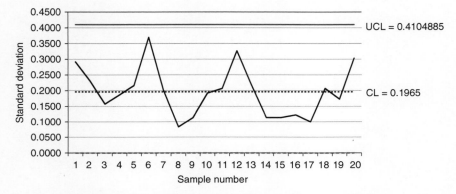

Figure 3.4 *S*-chart for net weight (in ounce) of a dry bleach product data.

In the chart if all the values of S falls within the control limits, then we proceed to construct the average chart. If larger proportion of values fall out of control, then we investigate the process for possible outliers or process special causes. The control limits are recalculated based on the subgroups considered as in-control points once ignoring the outliers or the points of the unusual causes after thorough investigation of the process movements. There is no need to update the limits without any sound evidence to change the control limits, or the control limits are not constructed using the enough data to draw conclusions from the results. Since all the points of the S table fall within the area of the in-control region, the $\overline{X}$ chart can easily be developed for this data set.

3.4.1 Construction of $\overline{X}$ Chart

Let n be the subgroup size and m be the total number of subgroups such that the variation within subgroups is smaller than between groups. Further let X_{ij} be the interested quality characteristic of the ith unit and the jth subgroup, where $i = 1, 2, 3, ..., n$ and $j = 1, 2, 3, ..., m$.

King (1954) was the first who developed the $\overline{X}$ control chart by adopting the grand average

$$\overline{\overline{X}} = \sum_{i=1}^{m} \overline{X}_i / m \tag{3.48}$$

$$\overline{\overline{X}} = \sum_{j=1}^{m} \sum_{i=1}^{n} X_{ij} / mn \tag{3.49}$$

and the average of the standard deviations of the subsamples $\overline{S} = \sum_{i=1}^{m} S_i / m$ to estimate the process mean and standard deviation, respectively. The three-sigma control limits using these estimates can be developed as

$$\text{UCL} = \text{Center line} + A_3 \overline{S} \tag{3.50}$$

$$\text{LCL} = \text{Center line} - A_3 \overline{S} \tag{3.51}$$

$$\text{UCL} = \overline{\overline{X}} + A_3 \overline{S} \tag{3.52}$$

$$\text{LCL} = \overline{\overline{X}} - A_3 \overline{S} \tag{3.53}$$

where A_3 is $3/c_4\sqrt{n}$ and a function that depends only on n, the subgroup size.

$$\text{If } \overline{S} = \sum_{i=1}^{m} S_i / m \tag{3.54}$$

and

$$S = \sqrt{\sum_{i=1}^{n} (X_i - \overline{X})^2 / n - 1} \tag{3.55}$$

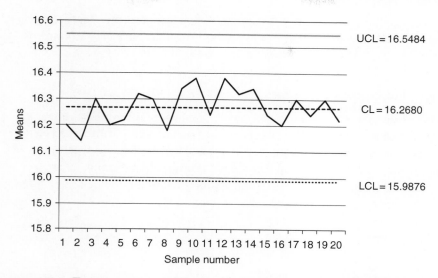

Figure 3.5 $\overline{X}$ chart using estimate $\overline{S}$ for net weight (in ounce) of a dry bleach product data.

then

$\overline{\overline{X}} = 16.2680$, $\overline{S} = 0.1965$, and $A_3 = 1.427$ (Figure 3.5).
Then UCL = 16.5484 and LCL = 15.9876.

The factors A_3, c_4, B_3, and B_4 (These are the constant values with respect to sample size and given in Appendix B.) we used above are based on the assumption that the process is normal, and we can use these constants even if the process is non-normal.

3.4.2 Normal and Non-normal Distributions for $\overline{X}$ and S-Charts

There is no theoretical justification for a not known to be normal distribution even approximately. Beginning with the notion of control chart, introduced by Shewhart, the methodology of control chart works very well for a wide range of distributions. As compared with the average and the dispersion chart, the latter is more sensitive to deviation from normality (Aijun, Sanyang, & Xiaojuan, 2016). In the light of the above fact, we can say that the mean chart is more reliable as compared with the dispersion chart even though it is built by using the dispersion chart.

3.5 Variance S^2-Charts

Traditionally, the Shewhart S^2-chart is recommended for large subgroup sizes using the estimate of σ^2 by considering the average value of the estimate (Raim et al., 1988; Serel & Moskowitz, 2008; Smith, 1994). Some quality control personnel also recommend this chart for small samples too. But small subgroup sizes can best be presented using the S-chart. The outliers are common in any data collection scenario. The main disadvantage of using the S^2-chart for such data leads to erroneous results as S^2 estimate is less robust against outliers.

3.5.1   Construction of S^2-Chart

Let n be the subgroup size and m be the total number of subgroups such that the variation within subgroups is smaller than between groups. Further let X_{ij} be the interested quality characteristic of the ith unit and the jth subgroup, where $i = 1, 2, 3, ..., n$ and $j = 1, 2, 3, ..., m$, and let $\overline{X}_j$ be the mean of a subgroup j of size n.

King (1954) was the first who developed the $\overline{X}$ control chart by adopting the grand average

$$X = \sum_{i=1}^{m} \overline{X}_i / m \tag{3.56}$$

$$X = \sum_{j=1}^{m} \sum_{i=1}^{n} X_{ij} / mn \tag{3.57}$$

and the average of the variances of the subsamples $\overline{S^2} = \sum_{i=1}^{m} S_i^2 / m$ to estimate the process mean and variance, respectively. The three-sigma control limits using these estimates can be developed as

$$\overline{S^2} = \sum_{i=1}^{m} S_i^2 / m \tag{3.58}$$

and

$$S^2 = \sum_{i=1}^{n} \left(X_i - \overline{X}\right)^2 / n - 1 \tag{3.59}$$

The UCL and LCL of the S^2 chart can be calculated as

$$\text{UCL} = \frac{\overline{S^2}}{n_j - 1} \chi^2_{\alpha/2, n_j - 1} \tag{3.60}$$

and

$$\text{LCL} = \frac{\overline{S^2}}{n_j - 1} \chi^2_{1 - \alpha/2, n_j - 1} \tag{3.61}$$

where $\chi^2_{1 - \alpha/2, n_j - 1}$ is the critical value of the chi-square distribution with K degrees of freedom.

For ease, we can define it as $S_{p,n}^2 = \chi^2_{p,n-1} / n - 1$.

Then the UCL and LCL can be defined as

$$\text{UCL} = \overline{S^2} * S_{\alpha/2, n}^2 \tag{3.62}$$

$$LCL = \overline{S^2} * S^2_{1-\alpha/2,n} \tag{3.63}$$

This control chart is defined with probability limits. As we know that when the process behavior does not follow the normal distribution, then probability limits are estimated instead of traditional three-sigma limits. These probability are estimated with the help of statistical software by fixing the false alarm probability at some acceptable level. In general, the probability of FAR is about 0.27% for the traditional three-sigma limits. The exact probability limits can be calculated if the exact distribution of the interested quality characteristic is known.

3.5.2 The Construction of Variance Chart for Neutrosophic Statistics

There are many situations in which the observation under study exhibits neutrosophic statistics. It is used in the situation when the observations are vague, incomplete, indeterminate, imprecise, or uncertain. For analysis of this type of data, a fuzzy approach is commonly applied. When observations consist of human subjectivity or any continuous variable involve human judgment for the observations, then the transmitted variability creates vagueness in the measurement system.

The customers always require the high-standard and high-quality goods and services to satisfy their desires (Aslam, Khan, & Khan, 2018). The control chart technique is used to monitor the special cause of variation in the process, and the standard of the product can only be maintained when the significant variation is timely detected and actively monitored. As we know, the location chart is used to monitor the variation between samples of size n, while the dispersion chart is used to monitor the variation within a sample of size n (van Zyl & van der Merwe, 2017). The variance chart is easy to apply as it consists of UCL, LCL, and a CL. A process is declared as out of control if any of the observation fall outside the UCL or LCL. There are many other signals for out-of-control process declaring, for instance, if two out of three observations or four out of five observations fall outside these limits. The performance evaluation of any chart can be judged through the ARL, which is defined as the average number of samples before any point in the control chart shows an out-of-control observation. The calculation of ARL is very common in the control chart literature and has been used by many authors of control chart literature.

Suppose a random sample is selected from the population or the sample from the indeterminate environment called the neutrosophic statistics. Let a neutrosophic random number $Y_{Ni}\epsilon\{Y_L, Y_U\}; i = 1, 2, 3, ..., n_N$, where Y_L is a precise part and Y_U is the imprecise part. The mean of precise and the imprecise population can be defined as $\mu_N = \sum_{i=1}^{N_N} Y_N/N_N; \mu_N\epsilon\{\mu_L,\mu_U\}$ and $\sigma_N^2 = \sum_{i=1}^{N_N}(Y_N-\mu_N)^2/N_N-1; \sigma_N^2 \epsilon\{\sigma_L^2,\sigma_U^2\}$, respectively, which denotes that the neutrosophic population variance with σ_L^2 is the variance of the precise part and σ_U^2 is the variance of the imprecise part. Let the neutrosophic sample mean of $Y_{Ni}\epsilon\{Y_L, Y_U\}; i = 1, 2, 3, ..., n_N$ be calculated as

$\overline{Y}_N = \sum_{i=1}^{N_N} Y_N/N_N; \overline{Y}_N \epsilon \{\overline{X}_L, \overline{X}_U\}$, and the neutrosophic sample variance of $Y_{Ni} \epsilon \{Y_L,$ $Y_U\}$; $i = 1, 2, 3, ..., n_N$ be calculated as $S_N^2 = \sum_{i=1}^{N_N}(Y_N - \overline{Y}_N)^2/N_N - 1; S_N^2 \epsilon \{S_L^2, S_U^2\}$.

This chart is specially developed for the variable inspection (measuring the quality of interest) to monitor the variance of the process.

The procedural steps of the variance neutrosophic statistic control chart can be described as follows:

Step I: In the first step we select a random sample of size n_N from the manufacturing process and calculate S_N^2.

Step II: If $LCL_N \leq S_N^2 \leq UCL_N$, then declare the process as in control, where $LCL_N \epsilon \{LCL_L, LCL_U\}$ and $UCL_N \epsilon \{UCL_L, UCL_U\}$ are defined as the neutrosophic interval control limits.

The neutrosophic interval control limits $LCL_N \epsilon \{LCL_L, LCL_U\}$ and $UCL_N \epsilon \{UCL_L, UCL_U\}$ can be constructed as

$$LCL_N = \sigma_N^2 - K_N \sqrt{2(\sigma_N^2)^2/(n_N - 1)}; \sigma_N^2 \epsilon \{\sigma_L^2, \sigma_L^2\}, K_N \epsilon \{K_L, K_U\} \qquad (3.64)$$

$$UCL_N = \sigma_N^2 + K_N \sqrt{2(\sigma_N^2)^2/(n_N - 1)}; \sigma_N^2 \epsilon \{\sigma_L^2, \sigma_L^2\}, K_N \epsilon \{K_L, K_U\} \qquad (3.65)$$

here $K_N \epsilon \{K_L, K_U\}$ is a neutrosophic statistic control chart coefficient and is computed through computer algorithm (Kenett, Zacks, & Amberti, 2013). It is to be noted that the variance control chart of the neutrosophic statistic is the generalization of the control chart through classical statistics (Yashchin, 1994). Therefore, when $K_L = K_U = K$ then this chart becomes the control chart under the classical statistics. The probability that the process is declared as out of control is calculated as

$$P_{outN}^{(0)} = P(S_N^2 \geq UCL_N) + P(S_N^2 \leq LCL_N); S_N^2 \epsilon \{S_L^2, S_U^2\} \qquad (3.66)$$

It can be observed that $(n_N - 1)S_N^2/\sigma_N^2; S_N^2 \epsilon \{S_L^2, S_L^2\}, \sigma_N^2 \epsilon \{S_L^2, S_L^2\}$ follows the neutrosophic chi-square distribution $\chi_N^2; \chi_N^2 \epsilon \{\chi_L^2, \chi_L^2\}$ with neutrosophic degrees of freedom $(n_N - 1)$; $n_N \epsilon \{n_L, n_U\}$ for the in-control process.

Let $G_N \epsilon \{G_L, G_U\}$ be the neutrosophic distribution function of $\chi_N^2 \epsilon \{\chi_L^2, \chi_U^2\}$. Then the in-control process will be

$$P(S_N^2 \geq UCL_n) = 1 - G_N \left(\frac{(n_N - 1)UCL_N}{\sigma_N^2} \right)$$

$$= 1 - G_N \left((n_N - 1)\left(1 + K_N \sqrt{2/(n_N - 1)} \right) \right);$$

$$n_N \in \{n_L, n_U\}, \quad K_N \in \{K_L, K_U\}. \qquad (3.67)$$

Likewise,

$$\left(S_N^2 \leq LCL_n\right) = G_N\left(\frac{(n_N - 1)UCL_N}{\sigma_N^2}\right)$$

$$= G_N\left((n_N - 1)\left(1 + K_N\sqrt{2/(n_N - 1)}\right)\right);$$

$$n_N \in \{n_L, n_U\}, \quad K_N \in \{K_L, K_U\}. \tag{3.68}$$

Now the probability of in-control process under the neutrosophic interval method can be written as

$$P_{outN}^{(0)} = 1 - G_N\left((n_N - 1)\left(1 + K_N\sqrt{2/(n_N - 1)}\right)\right)$$

$$+ G_N\left((n_N - 1)\left(1 - K_N\sqrt{2/(n_N - 1)}\right)\right);$$

$$n_N \epsilon \{n_L, n_U\}, K_N \epsilon \{K_L, K_U\} \tag{3.69}$$

The ARL is used commonly in the control chart literature to evaluate the performance of the control chart (Chakraborti, 2000). It is defined as the average number of samples until the process indicates an out-of-control signal. Thus ARL of the neutrosophic statistic denoted by NARL for variance chart using the neutrosophic interval method may be defined as

$$NARL_{0N} = \frac{1}{P_{outN}^{(0)}}; ARL_{0N} \epsilon \{ARL_{0L}, ARL_{0U}\} \tag{3.70}$$

Let the variance of the process has shifted to a new target value $\sigma_{1N}^2 = c\sigma_N^2; \sigma_N^2 \epsilon \{\sigma_L^2, \sigma_L^2\}$, where c denotes the shift in the process. Thus, the probability that the process has been shifted and declared as out of control, then

$$P_{outN}^{(1)} = P\left(S_N^2 \geq UCL_N | \sigma_{1N}^2\right) + P\left(S_N^2 \leq LCL_N | \sigma_{1N}^2\right); S_N^2 \epsilon \{S_L^2, S_L^2\} \tag{3.71}$$

Therefore, the probability of the shifted process for σ_{1N}^2 is defined as

$$P\left(S_N^2 \geq UCL_N | \sigma_{1N}^2\right) = 1 + G_N\left(\frac{(n_N - 1)}{c}\left(1 - K_N\sqrt{2/(n_N - 1)}\right)\right);$$

$$n_N \epsilon \{n_L, n_U\}, K_N \epsilon \{K_L, K_U\} \tag{3.72}$$

Similarly,

$$P\left(S_N^2 \leq LCL_N | \sigma_{1N}^2\right) = G_N\left(\frac{(n_N - 1)}{c}\left(1 - K_N\sqrt{2/(n_N - 1)}\right)\right);$$

$$n_N \epsilon \{n_L, n_U\}, K_N \epsilon \{K_L, K_U\} \tag{3.73}$$

Thus the probability of an out-of-control process under the neutrosophic interval method may be written as

$$P^{(1)}_{\text{outN}} = 1 + G_N\left(\frac{(n_N - 1)}{c}\left(1 - K_N\sqrt{2/(n_N - 1)}\right)\right)$$
$$+ G_N\left(\frac{(n_N - 1)}{c}\left(1 - K_N\sqrt{2/(n_N - 1)}\right)\right);$$

$$n_N \epsilon \{n_L, n_U\}, K_N \epsilon \{K_L, K_U\} \tag{3.74}$$

Then the NARL of the shifted process may be written as

$$\text{NARL}_{1N} = \frac{1}{P^{(1)}_{\text{outN}}}; \text{ARL}_{1N} \epsilon \{\text{ARL}_{1L}, \text{ARL}_{1U}\} \tag{3.75}$$

Using the abovementioned equations of the variance chart of the neutrosophic statistic, the NARL values have been estimated for different process settings. Table 3.4 generated for NARL when $n_N \epsilon \{3, 4\}$ with the in-control ARL fixed r_{0N} for 300 and 370. Table 3.5 generated for NARL when $n_N \epsilon \{4, 6\}$ with the in-control

Table 3.4 Neutrosophic ARL (NARL) values for chart parameters $n_N \epsilon \{3, 4\}$ and r_{0N} = 300, 370.

Parameters	r_{0N} = 300	r_{0N} = 370
n_N	[3, 4]	[3, 4]
k_N	[4.716, 4.784]	[4.921, 4.925]
c	NARL$_{1N}$	
1	[303.571, 482.268]	[372.916, 567.398]
1.1	[180.551, 257.614]	[217.685, 298.395]
1.2	[117.098, 153.218]	[138.998, 175.183]
1.3	[81.177, 98.952]	[95.096, 111.902]
1.4	[59.298, 68.163]	[68.685, 76.363]
1.5	[45.169, 49.433]	[51.809, 54.931]
1.6	[35.597, 37.374]	[40.482, 41.237]
1.7	[28.85, 29.239]	[32.562, 32.06]
1.8	[23.935, 23.534]	[26.833, 25.662]
1.9	[20.251, 19.399]	[22.567, 21.048]
2	[17.423, 16.317]	[19.311, 17.625]
3	[6.721, 5.593]	[7.198, 5.874]
4	[4.174, 3.353]	[4.394, 3.474]

Source: Aslam et al. (2018).

Table 3.5 Neutrosophic ARL (NARL) values for chart parameters $n_N \epsilon \{4, 6\}$ and r_{ON} = 300, 370.

Parameters	r_{ON} = 300		r_{ON} = 370
n_N	[4, 6]		[4, 6]
k_N	[4.37095, 4.38408]		[4.56277, 4.60236]
c		NARL$_{1N}$	
1	[300.03049, 490.72292]		[373.89462, 521.39944]
1.1	[167.75455, 236.62553]		[204.67051, 261.86897]
1.2	[103.63777, 130.01298]		[124.23386, 148.39858]
1.3	[69.11578, 78.91434]		[81.62574, 92.2282]
1.4	[48.93744, 51.76218]		[57.06262, 61.60346]
1.5	[36.34426, 36.10675]		[41.91335, 43.57537]
1.6	[28.05528, 26.46593]		[32.04372, 32.28357]
1.7	[22.35463, 20.20032]		[25.31664, 24.84157]
1.8	[18.28768, 15.94213]		[20.55525, 19.72459]
1.9	[15.29499, 12.93761]		[17.07608, 16.07855]
2	[13.03379, 10.74914]		[14.46377, 13.40067]
3	[4.84897, 3.5816]		[5.18032, 4.44486]
4	[3.02663, 2.21472]		[3.17374, 2.68331]

Source: Aslam et al. (2018).

ARL fixed r_{ON} for 300 and 370. Table 3.6 generated for NARL when $n_N \epsilon \{9, 10\}$ with the in-control ARL fixed r_{ON} for 300 and 370.

The variance control chart for the neutrosophic statistic S_N^2 has been given in Figure 3.6. The data have been generated from the neutrosophic normal distribution $\mu_N \epsilon \{0, 0\}$ and neutrosophic variance $\sigma_N^2 \epsilon \{4, 6.25\}$.

3.5.3 The Construction of Variance Chart for Repetitive Sampling

As discussed earlier the repetitive sampling scheme is better than the single sampling scheme in terms of the ARL. This sampling scheme was introduced by Sherman (1965). In repetitive group sampling scheme, which is different from double sampling scheme (He, Grigoryan, & Sigh, 2002; Hsu, 2004), the detection of the out-of-control process is based on selecting more than one sample from the production process. There are many situations where the distribution of the under-study variable is non-normal or not known. In general the abovementioned

Table 3.6 Neutrosophic ARL (NARL) values for chart parameters $n_N \epsilon \{9, 10\}$ and $r_{0N} = 300, 370$.

Parameters	$r_{0N} = 300$	$r_{0N} = 370$
n_N	[9, 10]	[9, 10]
k_N	[3.77774, 3.87857]	[3.90143, 3.92448]
c	NARL$_{1N}$	
1	[310.11015, 398.93766]	[374.85685, 429.14801]
1.1	[140.42988, 169.87263]	[166.07453, 181.17044]
1.2	[73.89048, 85.19126]	[85.81325, 90.2116]
1.3	[43.56058, 48.35266]	[49.82352, 50.89613]
1.4	[28.04099, 30.19755]	[31.65868, 31.6243]
1.5	[19.345, 20.33186]	[21.59801, 21.19929]
1.6	[14.10561, 14.53629]	[15.59629, 15.09889]
1.7	[10.75721, 10.90994]	[11.79341, 11.29453]
1.8	[8.51099, 8.52027]	[9.26122, 8.79481]
1.9	[6.94206, 6.87633]	[7.504, 7.0795]
2	[5.80835, 5.70385]	[6.24147, 5.85883]
3	[2.16011, 2.05723]	[2.24008, 2.0842]
4	[1.48785, 1.42243]	[1.51913, 1.43261]

Source: Aslam et al. (2018).

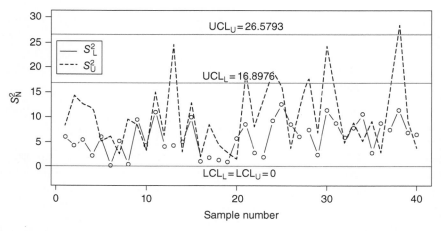

Figure 3.6 The S_N^2 control chart for the simulated data. *Source:* Aslam et al. (2018).

situation is replaced by using the nonparametric statistics, but the nonparametric tests are not so powerful and reliable for industrial statistics. Other solution to handle this situation is to use other non-normal distributions, for example, gamma distribution and beta distribution (Chen & Yeh, 2009).

Let X be a random variable with mean μ and variance σ^2. Further let a new random variable $Y_1, Y_2, Y_3, ..., Y_{\frac{n}{2}}$ defined as (Maedh et al., 2017)

$$Y_1 = \frac{(X_2 - X_1)^2}{2}$$

$$Y_2 = \frac{(X_4 - X_3)^2}{2}$$

$$Y_3 = \frac{(X_6 - X_5)^2}{2}$$

$$\vdots$$

$$Y_{\frac{n}{2}} = \frac{(X_n - X_{n-1})^2}{2}$$

with $E(Y_j) = \sigma^2$

Again let we define an indicator variable I as

$$I = \begin{cases} 1 & \text{if } Y_j > \sigma^2 \\ 0 & \text{otherwiswe} \end{cases} \tag{3.76}$$

Further let M be the total number of $Y_j > \sigma^2$, then this random variable $M = \sum_{j=1}^{\frac{n}{2}} I_j$ is a binomial random variable with parameters $\frac{n}{2}$ and P_0, where $P_0 = P(Y_j > \sigma^2)$.

Then the four control limits of the repetitive sampling scheme for this distribution can be written as

$$\text{LCL}_1 = \frac{n}{2}P_0 - K_1\sqrt{\frac{n}{2}P_0(1-P_0)} \tag{3.77}$$

$$\text{LCL}_2 = \frac{n}{2}P_0 - K_2\sqrt{\frac{n}{2}P_0(1-P_0)} \tag{3.78}$$

$$\text{UCL}_2 = \frac{n}{2}P_0 + K_2\sqrt{\frac{n}{2}P_0(1-P_0)} \tag{3.79}$$

$$\text{UCL}_1 = \frac{n}{2}P_0 + K_1\sqrt{\frac{n}{2}P_0(1-P_0)} \tag{3.80}$$

The operational procedure of the repetitive sampling scheme can be stated in the following steps:

Step I: Select a random sample of size n from the production process and compute the value of statistic M.

Step II: If $M_i > \text{UCL}_1$ or $M_i < \text{LCL}_1$, the process is declared as out of control.

Step III: If $\text{LCL}_2 < M_i < \text{UCL}_2$, the process is declared as in control. Otherwise go to the Step I.

To compute the values of ARL, we have to define the following probabilities. The probability of in control for the single sampling scheme is

$$P_{\text{ins}} = P(\text{LCL}_2 < M_i < \text{UCL}_2) \tag{3.81}$$

$$P_{\text{ins}} = \sum_{i=0}^{\frac{n}{2} + K_2\sqrt{\frac{n}{2}P_0(1-P_0)}} \binom{n/2}{i} P_0^i \left(1 - P_0^i\right)$$

$$- \sum_{i=0}^{\frac{n}{2} - K_2\sqrt{\frac{n}{2}P_0(1-P_0)}} \binom{n/2}{i} P_0^i \left(1 - P_0^i\right) \tag{3.82}$$

The probability of repetition denoted by P_{rep} is computed as

$$P_{\text{rep}} = P(\text{LCL}_1 < M_i < \text{LCL}_2) + P(\text{UCL}_2 < M_i < \text{UCL}_1) \tag{3.83}$$

$$P_{\text{rep}} = \sum_{i=0}^{\frac{n}{2} - K_2\sqrt{\frac{n}{2}P_0(1-P_0)}} \binom{n/2}{i} P_0^i \left(1 - P_0^i\right) - \sum_{i=0}^{\frac{n}{2} - K_1\sqrt{\frac{n}{2}P_0(1-P_0)}} \binom{n/2}{i}$$

$$P_0^i\left(1 - P_0^i\right) + \sum_{i=0}^{\frac{n}{2} + K_1\sqrt{\frac{n}{2}P_0(1-P_0)}} \binom{n/2}{i}$$

$$P_0^i\left(1 - P_0^i\right) - \sum_{i=0}^{\frac{n}{2} + K_2\sqrt{\frac{n}{2}P_0(1-P_0)}} \binom{n/2}{i} P_0^i \left(1 - P_0^i\right) \tag{3.84}$$

Then the probability of in control for the repetitive sampling scheme can be written as

$$P_{\text{in}} = \frac{P_{\text{ins}}}{1 - P_{\text{rep}}} \tag{3.85}$$

And the ARL of the in-control process can be written as

$$
\text{ARL}_0 = \cfrac{1}{1 - \left(\cfrac{\displaystyle\sum_{i=0}^{\frac{n}{2}+K_2\sqrt{\frac{n}{2}P_0(1-P_0)}} \binom{n/2}{i} P_0^i(1-P_0^i) - \displaystyle\sum_{i=0}^{\frac{n}{2}-K_2\sqrt{\frac{n}{2}P_0(1-P_0)}} \binom{n/2}{i} P_0^i(1-P_0^i)}{1 - \displaystyle\sum_{i=0}^{\frac{n}{2}-K_2\sqrt{\frac{n}{2}P_0(1-P_0)}} \binom{n/2}{i} P_0^i(1-P_0^i) - \displaystyle\sum_{i=0}^{\frac{n}{2}-K_1\sqrt{\frac{n}{2}P_0(1-P_0)}} \binom{n/2}{i} P_0^i(1-P_0^i) + \displaystyle\sum_{i=0}^{\frac{n}{2}+K_1\sqrt{\frac{n}{2}P_0(1-P_0)}} \binom{n/2}{i} P_0^i(1-P_0^i) - \displaystyle\sum_{i=0}^{\frac{n}{2}+K_2\sqrt{\frac{n}{2}P_0(1-P_0)}} \binom{n/2}{i} P_0^i(1-P_0^i)} \right)}
$$

$$(3.86)$$

The ARL of the shifted process denoted by ARL_1 when a shift of the form $P_1 \neq P_0$ for the single sampling scheme can be defined as

$$
P_{\text{ins},1} = P(\text{LCL}_2 < M_i < \text{UCL}_2 \mid P_1) \tag{3.87}
$$

$$
P_{\text{ins},1} = \sum_{i=0}^{\frac{n}{2}+K_2\sqrt{\frac{n}{2}P_0(1-P_0)}} \binom{n/2}{i} P_1^i(1-P_1^i) - \sum_{i=0}^{\frac{n}{2}-K_2\sqrt{\frac{n}{2}P_0(1-P_0)}} \binom{n/2}{i} P_1^i(1-P_1^i) \tag{3.88}
$$

The probability of repetitive sampling scheme is computed as

$$
P_{\text{rep},1} = P(\text{LCL}_1 < M_i < \text{LCL}_2) + P(\text{UCL}_2 < M_i < \text{UCL}_1) \mid P_1 \tag{3.89}
$$

$$
P_{\text{rep},1} = \sum_{i=0}^{\frac{n}{2}-K_2\sqrt{\frac{n}{2}P_0(1-P_0)}} \binom{n/2}{i} P_1^i(1-P_1^i) - \sum_{i=0}^{\frac{n}{2}-K_1\sqrt{\frac{n}{2}P_0(1-P_0)}} \binom{n/2}{i}
$$

$$
P_1^i(1-P_1^i) + \sum_{i=0}^{\frac{n}{2}+K_1\sqrt{\frac{n}{2}P_0(1-P_0)}} \binom{n/2}{i}
$$

$$
P_1^i(1-P_1^i) - \sum_{i=0}^{\frac{n}{2}+K_2\sqrt{\frac{n}{2}P_0(1-P_0)}} \binom{n/2}{i} P_1^i(1-P_1^i) \tag{3.90}
$$

Thus the probability of in control of the shifted process for the repetitive sampling is

$$
P_{\text{in1}} = \frac{P_{\text{ins},1}}{1 - P_{\text{rep},1}} \tag{3.91}
$$

Table 3.7 Average run length values for $ARL_0 = 300$ (Maedh et al., 2017).

P_0	0.35	0.40	0.45	0.50	0.55	0.60
n	120	90	71	62	28	77
K_1	3.2447	3.2015	2.8588	2.9566	2.9833	3.4946
K_2	0.7020	0.5833	2.3616	2.8929	1.0477	0.2467
Shifts				ARL_1		
1.00	300.16	302.03	300.25	300.58	300.63	302.44
0.95	295.30	279.44	209.46	220.09	173.95	86.54
0.90	143.86	134.19	110.55	115.02	96.99	22.43
0.85	55.74	52.35	56.41	57.76	54.05	6.34
0.80	20.71	19.64	29.75	30.29	30.40	2.32
0.75	7.83	7.51	16.41	16.82	17.32	1.32
0.70	3.27	3.18	9.49	9.91	10.04	1.07
0.65	1.72	1.69	5.77	6.18	5.97	1.02
0.60	1.21	1.21	3.71	4.08	3.70	1.00
0.50	1.01	1.01	1.85	2.12	1.74	1.00
0.40	1.00	1.00	1.23	1.37	1.18	1.00
0.30	1.00	1.00	1.04	1.09	1.03	1.00
0.20	1.00	1.00	1.00	1.01	1.00	1.00
0.10	1.00	1.00	1.00	1.00	1.00	1.00

The ARL values of the shifted can be computed using the following relation and given in Tables 3.7 and 3.8.

$$ARL_1 = \frac{1}{1 - P_{in1}} \tag{3.92}$$

$$ARL_1 = \frac{1}{1 - \frac{\sum_{i=0}^{\frac{n}{2}-K_2\sqrt{\frac{n}{4}P_0(1-P_0)}}\binom{n/2}{i}P_1^i(1-P_1^i) - \sum_{i=0}^{\frac{n}{2}-K_1\sqrt{\frac{n}{4}P_0(1-P_0)}}\binom{n/2}{i}P_1^i(1-P_1^i) + \sum_{i=0}^{\frac{n}{2}+K_1\sqrt{\frac{n}{4}P_0(1-P_0)}}\binom{n/2}{i}P_1^i(1-P_1^i) - \sum_{i=0}^{\frac{n}{2}+K_2\sqrt{\frac{n}{4}P_0(1-P_0)}}\binom{n/2}{i}P_1^i(1-P_1^i)}{\sum_{i=0}^{\frac{n}{2}+K_2\sqrt{\frac{n}{4}P_0(1-P_0)}}\binom{n/2}{i}P_1^i(1-P_1^i) - \sum_{i=0}^{\frac{n}{2}-K_2\sqrt{\frac{n}{4}P_0(1-P_0)}}\binom{n/2}{i}P_1^i(1-P_1^i)}} \tag{3.93}$$

Table 3.8 Average run length values for $ARL_0 = 370$ (Maedh et al., 2017).

P_0	0.35	0.40	0.45	0.50	0.55	0.60
n	82	24	46	52	46	34
K_1	2.8852	3.2317	2.8231	2.9661	3.0680	3.0462
K_2	2.8628	1.6889	1.6034	1.6767	1.7808	2.9018
Shifts			ARL_1			
1.00	370.02	370.90	370.94	371.69	370.94	371.66
0.95	366.59	273.74	301.52	279.10	270.86	182.74
0.90	238.12	194.24	183.59	149.29	139.56	92.56
0.85	132.13	135.94	101.35	74.04	67.20	49.57
0.80	72.29	94.89	55.07	37.02	32.86	28.09
0.75	40.57	66.35	30.13	19.00	16.59	16.78
0.70	23.56	46.53	16.71	10.07	8.72	10.54
0.65	14.20	32.73	9.44	5.58	4.83	6.94
0.60	8.90	23.08	5.49	3.29	2.89	4.78
0.50	3.94	11.54	2.20	1.54	1.43	2.57
0.40	2.09	5.82	1.27	1.10	1.08	1.63
0.30	1.35	3.00	1.04	1.01	1.01	1.22
0.20	1.07	1.68	1.00	1.00	1.00	1.05
0.10	1.00	1.13	1.00	1.00	1.00	1.00

References

Ahmad, L., Aslam, M., & Jun, C.-H. (2014a). Coal quality monitoring with improved control charts. *European Journal of Scientific Research*, *125*(2), 427–434.

Ahmad, L., Aslam, M., & Jun, C.-H. (2014b). Designing of $\overline{X}$ control charts based on process capability index using repetitive sampling. *Transactions of the Institute of Measurement and Control*, *36*(3), 367–374.

Ahmad, L., Aslam, M., & Jun, C.-H. (2016). The design of a new repetitive sampling control chart based on process capability index. *Transactions of the Institute of Measurement and Control*, *38*(8), 971–980.

Aijun, Y., Sanyang, L., & Xiaojuan, D. (2016). Variables two stage sampling plans based on the coefficient of variation. *Journal of Advanced Mechanical Design, Systems, and Manufacturing*, *10*(1), 1–12, JAMDSM0002-JAMDSM0002

Allen, T. T. (2006). *Introduction to engineering statistics and six sigma*. London: Springer-Verlag.

Amiri, F., Noghondarian, K., & Noorossana, R. (2014). Economic-statistical design of adaptive $\overline{X}$ control chart: A Taguchi loss function approach. *Scientia Iranica. Transaction E, Industrial Engineering, 21*(3), 1096.

Aslam, M., Khan, N., & Khan, M. (2018). Monitoring the variability in the process using neutrosophic statistical interval method. *Symmetry, 10*(11), 562.

Azam, M., Ahmad, L., & Aslam, M. (2016). Design of $\overline{X}$ chart for Burr distribution under the repetitive sampling. *Science International, 28*(4), 3265–3271.

Bowker, A. H., & Goode, H. P. (1952). *Sampling inspection by variables.* New York: McGraw-Hill.

Castagliola, P., Celano, G., & Psarakis, S. (2011). Monitoring the coefficient of variation using EWMA charts. *Journal of Quality Technology, 43*(3), 249.

Chakraborti, S. (2000). Run length, average run length and false alarm rate of Shewhart $\overline{X}$ chart: exact derivations by conditioning. *Communications in Statistics-Simulation and Computation, 29*(1), 61–81.

Chakraborti, S., Human, S. W., & Graham, M. A. (2008). Phase I statistical process control charts: An overview and some results. *Quality Engineering, 21*(1), 52–62. doi:10.1080/08982110802445561

Chandara, M. J. (2001). *Statistical quality control.* Boca Raton, FL: CRC Press LLC.

Chang, T., & Gan, F. (2004). Shewhart charts for monitoring the variance components. *Journal of Quality Technology, 36*(3), 293.

Chao-Wen, L., & Reynolds, M. R., Jr. (1999). EWMA control charts for monitoring the mean of autocorrelated processes. *Journal of Quality Technology, 31*(2), 166.

Chen, F., & Yeh, C.-H. (2009). Economic statistical design of non-uniform sampling scheme X bar control charts under non-normality and gamma shock using genetic algorithm. *Expert Systems with Applications, 36*(5), 9488–9497.

Chen, F.-L., & Yeh, C.-H. (2011). Economic statistical design for $\overline{x}$ control charts under non-normal distributed data with Weibull in-control time. *Journal of the Operational Research Society, 62*(4), 750–759.

Chen, G., & Cheng, S. W. (1998). Max chart: Combining $\overline{X}$ chart and S chart. *Statistica Sinica, 8*, 263–271.

Gan, F. (1995). Joint monitoring of process mean and variance using exponentially weighted moving average control charts. *Technometrics, 37*(4), 446–453.

Hawkins, D. M., & Olwell, D. H. (2012). *Cumulative sum charts and charting for quality improvement.* New York: Springer Science & Business Media.

He, D., Grigoryan, A., & Sigh, M. (2002). Design of double-and triple-sampling $\overline{X}$ control charts using genetic algorithms. *International Journal of Production Research, 40*(6), 1387–1404.

Hsu, L.-F. (2004). Note on "design of double-and triple-sampling $\overline{X}$ control charts using genetic algorithms.". *International Journal of Production Research, 42*(5), 1043–1047.

Kang, C. W., Lee, M. S., Seong, Y. J., & Hawkins, D. M. (2007). A control chart for the coefficient of variation. *Journal of Quality Technology, 39*(2), 151.

Kenett, R., Zacks, S., & Amberti, D. (2013). *Modern industrial statistics: With applications in R, MINITAB and JMP.* Chichester, UK: Wiley.

Khoo, M. B. (2005). A modified S chart for the process variance. *Quality Engineering, 17*(4), 567–577.

King, E. P. (1954). Probability limits for the average chart when process standards are unspecified. *Industrial Quality Control, 10*(6), 62–64.

Koo, T.-Y., & Case, K. E. (1990). Economic design of $\bar{X}$ control charts for use in monitoring continuous flow processes. *The International Journal of Production Research, 28*(11), 2001–2011.

Kruger, U., & Xie, L. (2012). *Advances in statistical monitoring of complex multivariate processes: With applications in industrial process control.* Chichester, UK: Wiley.

Liu, H.-R., Chou, C.-Y., & Chen, C.-H. (2002). Minimum-loss design of $\bar{x}$ charts for correlated data. *Journal of Loss Prevention in the Process Industries, 15*(6), 405–411.

Lu, C.-W., & Reynolds, M. R. (1999). Control charts for monitoring the mean and variance of autocorrelated processes. *Journal of Quality Technology, 31*(3), 259–274.

Maedh, A. A., Ahmad, L., & Khan, K. (2017). A new control chart for monitoring process variance under repetitive group sampling scheme. *Journal of Computational and Theoretical Nanoscience, 14*(12), 5704–5710.

Mason, R. L., & Young, J. C. (2002). *Multivariate statistical process control with industrial applications* (Vol. 9). Philadelphia, PA: Siam.

Montgomery, C. D. (2009). *Introduction to statistical quality control* (6th ed.). New York: Wiley

Oakland, J. S. (2008). *Statistical process control* (6th ed.). Oxford, UK: Butterworth-Heinemann.

Pearn, W. L., & Kotz, S. (2006). *Encyclopedia and handbook of process capability indices: A comprehensive exposition of quality control measures series on quality, reliability and engineering statistics* (Vol. 12). Singapore: World Scientific Publishing Co. Pte. Ltd.

Prajapati, D., & Mahapatra, P. (2009). A simple and effective R chart to monitor the process variance. *International Journal of Quality & Reliability Management, 26*(5), 497–512.

Rahim, M., & Banerjee, P. (1993). A generalized model for the economic design of $\bar{x}$ control charts for production systems with increasing failure rate and early replacement. *Naval Research Logistics, 40*(6), 787–809.

Raim, M., Lashkari, R., & Banerjee, P. K. (1988). Joint economic design of mean and variance control charts. *Engineering Optimization, 14*(1), 65–78.

Schilling, E. G., & Neubauer, D. V. (2009). *Acceptance Sampling in Quality Control.* Boca Raton, FL: CRC Press.

Serel, D. A., & Moskowitz, H. (2008). Joint economic design of EWMA control charts for mean and variance. *European Journal of Operational Research, 184*(1), 157–168.

Sherman, R. E. (1965). Design and evaluation of a repetitive group sampling plan. *Technometrics, 7*(1), 11–21.

Shewhart, W. A. (1931). *Economic Control of Quality of Manufactured Product*. New York: D Van Nostrand Co.

Smith, A. E. (1994). $\overline{X}$ and R control chart interpretation using neural computing. *The International Journal of Production Research, 32*(2), 309–320.

Subramani, J., & Balamurali, S. (2012). Control charts for variables with specified process capabilities indices. *International Journal of Probability and Statistics, 1*(4), 101–110.

Tong, Y., & Chen, Q. (1991). Sampling inspection by variables for coefficient of variation. *Theory and Applied Probability, 3*, 315–327.

Trietsch, D. (1998). *Statistical quality control: A loss minimization approach* (Vol. 10). Singapore: World Scientific Publishing Co. Pte Ltd.

van Zyl, R., & van der Merwe, A. J. (2017). A Bayesian control chart for a common coefficient of variation. *Communications in Statistics – Theory and Methods, 46*(12), 5795–5811. doi:10.1080/03610926.2015.1112914

Xie, M., Goh, T. N., & Kuralmani, V. (2012). *Statistical models and control charts for high-quality processes*. New York: Springer Science & Business Media.

Yan, A., Liu, S., & Azam, M. (2017). Designing a multiple state repetitive group sampling plan based on the coefficient of variation. *Communications in Statistics – Simulation and Computation, 46*(9), 7154–7165.

Yashchin, E. (1994). Monitoring variance components. *Technometrics, 36*(4), 379–393.

4

Control Chart for Multiple Dependent State Sampling

4.1 Introduction

Multiple dependent state (MDS) sampling was originally proposed by Wortham and Baker (1976). Later on Balamurali and Jun (2007) designed sampling plan for the MDS sampling using the variable data. The MDS sampling scheme is considered as more flexible and more efficient than the single sampling plan in sample size required for the inspection/testing of the lot of the product. The MDS sampling utilizes the current lot information and previous accepted lot information to make the decision about the lot of the product. This sampling scheme has been widely used in the area of acceptance sampling plans for the lot sentencing purpose, for example (Aslam, Yen, Chang, & Jun, 2013; Balamurali, Jeyadurga, & Usha, 2017). Aslam, Nazir, and Jun (2015) originally introduced the area of control chart and proposed attribute control chart using the MDS sampling and showed the efficiency of the attribute control chart over the traditional Shewhart attribute control chart in terms of average run length (ARL). The use of the MDS sampling in the area of control chart has increased the sensitivity of the control charts to detect a small shift in the manufacturing process. For decision making, it uses the current subgroup information and previous subgroup information to make the decision about the state of the process.

In this chapter, we will discuss attribute and variable control charts using the MDS sampling. We will also discuss control chart using process capability index for the normal distribution and non-normal distribution using the MDS sampling.

4.2 Attribute Charts Using MDS Sampling

In this section, we present the design of attribute control charts using the MDS sampling. The attribute control charts are widely used for the monitoring of

Introduction to Statistical Process Control, First Edition. Muhammad Aslam, Aamir Saghir, and Liaquat Ahmad.

nonconforming items. The attribute control charts are applied when the data is obtained from the counting/no-go-no process.

4.2.1 The *np*-Control Chart

Aslam, Nazir, et al. (2015) introduced the *np*-control chart using the MDS sampling. Let p be the probability of defective/nonconforming items. It is assumed that p will remain same for each subgroup size n. So the binomial distribution can be used to design the *np*-control chart. Following the study of Montgomery (2009), the upper control limit (UCL) and the lower control limit (LCL) of the traditional Shewhart *np*-control chart are given as

$$\text{UCL} = np + 3\sqrt{np(1-p)} \tag{4.1}$$

$$\text{LCL} = np - 3\sqrt{np(1-p)} \tag{4.2}$$

The process is declared to be in control state if the plotting statistic which number of defective D is within the UCL and LCL. If $D < \text{LCL}$ or $D > \text{UCL}$, then the process is said to be declared to be out of control. Suppose that p_0 shows the in-control fraction nonconforming, when p_0 is small, the LCL is set to 0 and unable to detect shift in p.

The *np*-control chart using the MDS sampling has two UCLs and two inner control limits (Aslam, Nazir, et al., 2015).

Step I: Establish the outer control limits of UCL_1 and LCL_1 as well as the inner control limits of UCL_2 and LCL_2 using the data for the in-control process.

Step II: Select a random sample of size n from the production process at each subgroup and count the number of defectives D.

Step III: Declare the process as in control if $\text{LCL}_2 \leq D \leq \text{UCL}_2$. Declare the process to be out of control if $D \geq \text{UCL}_1$ or $D \leq \text{LCL}_1$. Otherwise, go to Step IV.

Step IV: Declare the process as in control if i proceeding subgroups declared the process as in control. Otherwise, declare the process to be out of control.

The two inner and outer control limits for the *np*-control chart are given by

$$\text{UCL}_1 = np_0 + k_1\sqrt{np_0(1-p_0)} \tag{4.3}$$

$$\text{LCL}_1 = \max\left[0, np_0 - k_1\sqrt{np_0(1-p_0)}\right] \tag{4.4}$$

$$\text{UCL}_2 = np_0 + k_2\sqrt{np_0(1-p_0)} \tag{4.5}$$

$$\text{LCL}_2 = \max\left[0, np_0 - k_2\sqrt{np_0(1-p_0)}\right] \tag{4.6}$$

where k_1 and k_2 are two control chart coefficients and will be determined by the simulation. In practice, p_0 is unknown and can be estimated from preliminary sample data. The probability of declaring the process in control is stated by

$$P_{in} = P(LCL_2 \leq D \leq UCL_2) + \{P(LCL_1 < D < LCL_2)$$

$$+ P(UCL_2 < D < UCL_1)\}\{P(LCL_2 \leq D \leq UCL_2)\}^i \qquad (4.7)$$

Let

$$A_1 = P(LCL_2 \leq D \leq UCL_2) = \sum_{d = |LCL_2| + 1}^{|UCL_2|} \binom{n}{d} p^d (1-p)^{n-d} d = 0, 1, ..., n$$
$$(4.8)$$

$$A_2 = P(LCL_1 < D < LCL_2) = \sum_{d = |LCL_1| + 1}^{|LCL_2|} \binom{n}{d} p^d (1-p)^{n-d} d = 0, 1, ..., n$$
$$(4.9)$$

$$A_3 = P(UCL_2 < D < UCL_1) = \sum_{d = |UCL_2| + 1}^{|UCL_1|} \binom{n}{d} p^d (1-p)^{n-d} d = 0, 1, ..., n$$
$$(4.10)$$

So the P_{in} in given in Eq. (4.7) is given by

$$P_{in} = A_1 + \{A_2 + A_3\}\{A_1\}^i \qquad (4.11)$$

Let p_0 be the percent defective for the in-control process, so the probability of in control for np chart using the MDS sampling is given by

$$P_{in}^0 = A_1^0 + \{A_2^0 + A_3^0\}\{A_1^0\}^i \qquad (4.12)$$

where

$$A_1^0 = P(LCL_2 \leq D \leq UCL_2) = \sum_{d = |LCL_2| + 1}^{|UCL_2|} \binom{n}{d} p_0^d (1-p_0)^{n-d} \qquad (4.13)$$

$$A_2^0 = P(LCL_1 < D < LCL_2) = \sum_{d = |LCL_1| + 1}^{|LCL_2|} \binom{n}{d} p_0^d (1-p_0)^{n-d} \qquad (4.14)$$

$$A_3^0 = P(UCL_2 < D < UCL_1) = \sum_{d = |UCL_2| + 1}^{|UCL_1|} \binom{n}{d} p_0^d (1-p_0)^{n-d} \qquad (4.15)$$

The ARL at p_0 is given by

$$ARL_0 = \frac{1}{1 - P_{in}^0} \qquad (4.16)$$

Similarly, the necessary measures for the shifted process at $p_1 = p_0 + cp_0 = (1 + c)p_0$, where c is a constant, are given as

$$P_{in}^1 = A_1^1 + \{A_2^1 + A_3^1\}\{A_1^1\}^i \qquad (4.17)$$

where

$$A_1^1 = P(\text{LCL}_2 \le D \le \text{UCL}_2) = \sum_{d = |\text{LCL}_2| + 1}^{|\text{UCL}_2|} \binom{n}{d} p_1^d (1 - p_1)^{n-d} \tag{4.18}$$

$$A_2^1 = P(\text{LCL}_1 < D < \text{LCL}_2) = \sum_{d = |\text{LCL}_1| + 1}^{|\text{LCL}_2|} \binom{n}{d} p_1^d (1 - p_1)^{n-d} \tag{4.19}$$

$$A_3^0 = P(\text{UCL}_2 < D < \text{UCL}_1) = \sum_{d = |\text{UCL}_2| + 1}^{|\text{UCL}_1|} \binom{n}{d} p_1^d (1 - p_1)^{n-d} \tag{4.20}$$

The ARL for the shifted process is reported as

$$\text{ARL}_1 = \frac{1}{1 - P_{\text{in}}^1} \tag{4.21}$$

Tables 4.1 and 4.2 are reported for specified ARLs 300 and 370, various n, and processing in-control subgroup i. It can be noted that ARL_1 decreases when i increases from 2 to 3.

An example is selected about the packing of frozen orange juice concentrate. As mentioned by Montgomery (2009), "Frozen orange juice concentrate is packed in 6-oz cardboard cans. These cans are formed on a machine by spinning them from cardboard stock and attaching a metal bottom panel. By inspection of a can, it may determine whether, when filled, it could possibly leak either on the side seam or around the bottom joint. Such a nonconforming can have an improper seal on either the side seam or the bottom panel." For this data $n = 50$ each of size 30. From Table 4.2, $k_1 = 3.850,548$, $k_2 = 2.145,123$, and $i = 3$.

The necessary computations are shown as

$$\bar{p} = \frac{\sum_{i=1}^{m} D_i}{mn} = \frac{347}{1500} = 0.2313$$

$$\text{UCL}_1 = n\bar{p} + k_1 \sqrt{n\bar{p}(1 - \bar{p})} = 23.045$$

$$\text{LCL}_1 = \max\left[0, n\bar{p} - k_1 \sqrt{n\bar{p}(1 - \bar{p})}\right] = 1.42756$$

$$\text{UCL}_2 = n\bar{p} + k_2 \sqrt{n\bar{p}(1 - \bar{p})} = 17.96$$

$$\text{LCL}_2 = \max\left[0, n\bar{p} - k_2 \sqrt{n\bar{p}(1 - \bar{p})}\right] = 5.169$$

Figure 4.1 shows that although the data is obtained from the in-control process, some points are within the in-decision area and out of control. The control chart using the MDS sampling indicates that some action should be taken to bring back the process to in-control state. While the existing control chart for the same data given in Montgomery (2009) shows, the process is in control state, and no action is needed. The efficiency of the control chart using MDS is also shown using the simulated data in Figure 4.2, whereas the efficiency of the existing chart is shown in

Table 4.1 ARL_1 of the proposed control chart for $r_0 = 370$ (Aslam, Nazir, et al., 2015).

c	$n = 40$ $k_1 = 3.067536$ $k_2 = 2.304818$ $i = 2$	$n = 50$ $k_1 = 3.102263$ $k_2 = 2.559945$ $i = 3$	$n = 55$ $k_1 = 4.167968$ $k_2 = 2.180153$ $i = 3$	$n = 65$ $k_1 = 3.096783$ $k_2 = 2.687494$ $i = 3$	$n = 75$ $k_1 = 4.141027$ $k_2 = 2.160161$ $i = 2$	$n = 85$ $k_1 = 3.586033$ $k_2 = 2.185251$ $i = 3$
0	371.17	373.02	376.65	372.51	370.14	371.66
0.1	99.44	88.85	58.25	68.68	41.32	31.59
0.2	15.43	15.43	6.96	10.43	4.55	3.47
0.3	4.07	4.07	2.07	2.82	1.52	1.37
0.4	1.78	1.78	1.23	1.41	1.07	1.05
0.5	1.19	1.19	1.04	1.07	1.00	1.00
0.6	1.03	1.03	1.00	1.00	1.00	1.00
0.7	1.00	1.00	1.00	1.00	1.00	1.00
0.8	1.00	1.00	1.00	1.00	1.00	1.00
0.9	1.00	1.00	1.00	1.00	1.00	1.00
1	1.00	1.00	1.00	1.00	1.00	1.00

Table 4.2 ARL_1 of the proposed control chart for $r_0 = 300$ (Aslam, Nazir, et al., 2015).

	$n = 10$	$n = 35$	$n = 45$	$n = 50$	$n = 60$	$n = 80$	$n = 150$
	$k_1 = 3.04250$	$k_1 = 4.15763$	$k_1 = 3.2940$	$k_1 = 3.85054$	$k_1 = 3.2169$	$k_1 = 3.67668$	$k_1 = 3.968$
	$k_2 = 2.474011$	$k_2 = 2.163697$	$k_2 = 2.03049$	$k_2 = 2.145123$	$k_2 = 2.1084$	$k_2 = 2.2258$	$k_2 = 1.965$
c	$i = 3$	$i = 2$	$i = 2$	$i = 3$	$i = 3$	$i = 3$	$i = 2$
0	313.95	302.36	305.85	311.35	316.21	300.79	300.99
0.1	204.54	86.66	66.36	56.56	46.85	29.95	12.79
0.2	84.02	14.33	9.71	7.44	5.93	3.54	1.68
0.3	32.42	3.84	2.71	2.22	1.90	1.40	1.048
0.4	13.35	1.73	1.39	1.28	1.19	1.06	1.00
0.5	6.13	1.18	1.08	1.05	1.03	1.00	1.00
0.6	3.22	1.03	1.01	1.01	1.00	1.00	1.00
0.7	1.95	1.00	1.00	1.00	1.00	1.00	1.00
0.8	1.37	1.00	1.00	1.00	1.00	1.00	1.00
0.9	1.09	1.00	1.00	1.00	1.00	1.00	1.00
1	1.00	1.00	1.00	1.00	1.00	1.00	1.00

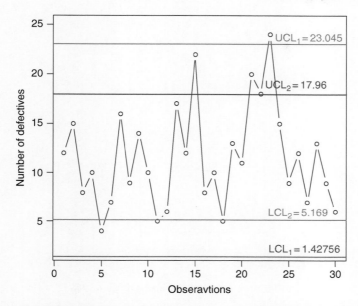

Figure 4.1 Proposed control chart for the example (Aslam, Nazir, et al., 2015).

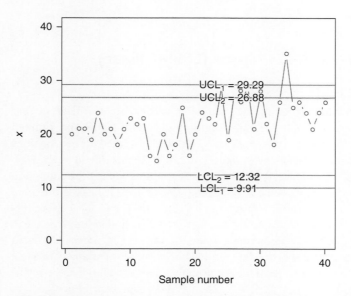

Figure 4.2 Proposed control chart for simulated data (Aslam, Nazir, et al., 2015).

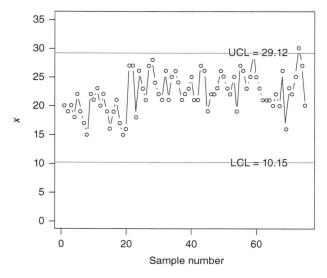

Figure 4.3 Existing control chart for simulated data (Aslam, Nazir, et al., 2015).

Figure 4.3. The control chart using MDS sampling in Figure 4.2 indicates the shift in the process, while the Shewhart control chart does not detect this shift in the process.

4.3 Conway–Maxwell–Poisson (COM–Poisson) Distribution

The Poisson distribution is commonly used when it is expected to have smaller number of defectives per unit or item. According to Sellers, Borle, and Shmueli (2012), this distribution has many applications and the best fitted in the situations where it is assumed that mean and variance are same. As mentioned by Conway and Maxwell (1962), in practice, there are several situations where the assumption of the identical mean and variance is violated and recommended to use the more flexible distribution called the COM–Poisson distribution. The COM–Poisson distribution can be applied on count data when the variance is smaller than the mean which is known as under-dispersed data, when the variance is larger than mean which is known as over-dispersed data, and when mean and variance are same which is known as equi-dispersed data. Some applications of COM–Poisson distribution can be seen in Sellers et al. (2012). The COM–Poisson distribution is more efficient than generalized Poisson (GP) distribution and two-parameter GP distribution derived by del Castillo and Pérez-Casany (2005) and Olkin (1992), respectively. Famoye (1994) proposed control chart using GP distribution.

The probability mass function of the COM–Poisson distribution having scale parameter $\mu(>0)$ and dispersion parameter $v(\geq 0)$ proposed by Conway and Maxwell (1962) is given by

$$P(X_t = x/\mu, v) = \frac{\mu^x}{(x!)^v z(\mu, v)}, \quad \text{for } x = 0, 1, 2, \dots \qquad (4.22)$$

where $z(\mu, v) = \sum_{s=0}^{\infty} \frac{\mu^s}{(s!)^v}$ is normalizing constant. The Poisson distribution when $v = 0$, $\mu < 1$, the geometric distribution when $v \to \infty$ with probability $\mu/(1 + \mu)$, and Bernoulli distribution when $v > 1$ are the special cases of the Poisson distribution. Aslam, Ahmad, Jun, and Arif (2016) proposed the control chart for the COM–Poisson distribution using the MDS sampling. The operational process according to Aslam et al. (2016) is stated as follows:

Step I: Select an item from the production process at time t and count the number of nonconformities, say X_t. Compute exponentially weighted moving average statistic at time t as

$$Z_t = \lambda X_t + (1 - \lambda)Z_{t-1} \qquad (4.23)$$

where λ is smoothing constant.

Step II: The process is declared to be in control state if $\text{LCL}_2 \leq Z_t \leq \text{UCL}_2$. Otherwise, go to Step III.

Step III: The process is said to be in control if i proceeding subgroups are in control using Step II.

The mean and variance of statistic Z_t are given by

$$\mu_Z = \mu_0^{1/v} - \frac{v - 1}{2v} \qquad (4.24)$$

$$\sigma_Z^2 = \frac{\lambda}{2 - \lambda} \left[\frac{\mu_0^{1/v}}{v} \right] \qquad (4.25)$$

The four control limits using for the control chart are given by

$$\text{LCL}_1 = \left(\mu_0^{1/v} - \frac{v - 1}{2v} \right) - k_1 \sqrt{\frac{\lambda}{2 - \lambda} \left[\frac{\mu_0^{1/v}}{v} \right]} \qquad (4.26)$$

$$\text{UCL}_1 = \left(\mu_0^{1/v} - \frac{v - 1}{2v} \right) + k_1 \sqrt{\frac{\lambda}{2 - \lambda} \left[\frac{\mu_0^{1/v}}{v} \right]} \qquad (4.27)$$

$$\mathrm{LCL}_2 = \left(\mu_0^{1/v} - \frac{v-1}{2v}\right) - k_2\sqrt{\frac{\lambda}{2-\lambda}\left[\frac{\mu_0^{1/v}}{v}\right]} \tag{4.28}$$

$$\mathrm{UCL}_2 = \left(\mu_0^{1/v} - \frac{v-1}{2v}\right) + k_2\sqrt{\frac{\lambda}{2-\lambda}\left[\frac{\mu_0^{1/v}}{v}\right]} \tag{4.29}$$

The probability of in-control, say P_{in}^0, is given by

$$P_{\mathrm{in}}^0 = (\Phi(k_2) - \Phi(-k_2)) + \{(\Phi(-k_2) - \Phi(-k_1) + (\Phi(k_1)$$
$$- \Phi(k_2)))\}\{(\Phi(k_2) - \Phi(-k_2))\}^i \tag{4.30}$$

The ARL when the process is at μ_0 (in-control scale parameter) is given by

$$\mathrm{ARL}_0 = \frac{1}{1 - \left((\Phi(k_2) - \Phi(-k_2)) + \{(\Phi(-k_2) - \Phi(-k_1) + (\Phi(k_1) - \Phi(k_2)))\}\{(\Phi(k_2) - \Phi(-k_2))\}^i\right)} \tag{4.31}$$

The probability of out of control at $\mu = c * \mu_0$ is given by

$$P_{\mathrm{in}}^1 = A_1^1 + \{A_2^1 + A_3^1\}\{A_1^1\}^i \tag{4.32}$$

where

$$A_1^1 = P(\mathrm{LCL}_2 \le Z_t \le \mathrm{UCL}_2 \mid \mu = c * \mu_0)$$

$$= \Phi\left(\frac{(1 - c^{1/v})\mu_0^{1/v} + k_2\sqrt{\frac{\lambda}{2-\lambda}\left[\frac{\mu_0^{1/v}}{v}\right]}}{\sqrt{\frac{\lambda}{2-\lambda}\left[\frac{\mu_0^{1/v}c^{1/v}}{v}\right]}}\right)$$

$$- \Phi\left(\frac{(1 - c^{1/v})\mu_0^{1/v} - k_2\sqrt{\frac{\lambda}{2-\lambda}\left[\frac{\mu_0^{1/v}}{v}\right]}}{\sqrt{\frac{\lambda}{2-\lambda}\left[\frac{\mu_0^{1/v}c^{1/v}}{v}\right]}}\right)$$

$$A_2^1 = P(\mathrm{LCL}_1 < Z_t < \mathrm{LCL}_2 \mid \mu = c * \mu_0)$$

$$= \Phi\left(\frac{(1 - c^{1/v})\mu_0^{1/v} - k_2\sqrt{\dfrac{\lambda}{2-\lambda}\left[\dfrac{\mu_0^{1/v}}{v}\right]}}{\sqrt{\dfrac{\lambda}{2-\lambda}\left[\dfrac{\mu_0^{1/v}c^{1/v}}{v}\right]}}\right)$$

$$- \Phi\left(\frac{(1 - c^{1/v})\mu_0^{1/v} - k_1\sqrt{\dfrac{\lambda}{2-\lambda}\left[\dfrac{\mu_0^{1/v}}{v}\right]}}{\sqrt{\dfrac{\lambda}{2-\lambda}\left[\dfrac{\mu_0^{1/v}c^{1/v}}{v}\right]}}\right)$$

$$A_3^1 = P(\mathrm{UCL}_2 < Z_t < \mathrm{UCL}_1 \mid \mu = c * \mu_0)$$

$$= \Phi\left(\frac{(1 - c^{1/v})\mu_0^{1/v} + k_1\sqrt{\dfrac{\lambda}{2-\lambda}\left[\dfrac{\mu_0^{1/v}}{v}\right]}}{\sqrt{\dfrac{\lambda}{2-\lambda}\left[\dfrac{\mu_0^{1/v}c^{1/v}}{v}\right]}}\right)$$

$$- \Phi\left(\frac{(1 - c^{1/v})\mu_0^{1/v} + k_2\sqrt{\dfrac{\lambda}{2-\lambda}\left[\dfrac{\mu_0^{1/v}}{v}\right]}}{\sqrt{\dfrac{\lambda}{2-\lambda}\left[\dfrac{\mu_0^{1/v}c^{1/v}}{v}\right]}}\right)$$

The ARL for the shifted process is given by

$$\mathrm{ARL}_1 = \frac{1}{1 - \left(A_1^1 + \{A_2^1 + A_3^1\}\{A_1^1\}^i\right)} \tag{4.33}$$

Tables 4.3–4.5 are selected from Aslam et al. (2016), which clearly indicate that ARLs decrease as μ_0 increases. Table 4.6 shows the comparison between Aslam et al. (2016) control chart and Saghir and Lin (2014) control chart. It is clear from the Table 4.6 that COM–Poisson distribution using MDS sampling performs better than Saghir and Lin (2014) control chart in terms of ARLs. Similarly, the performance of the Aslam et al. (2016) control chart is also shown using the simulation data where 30 observations are generated from in-control parameters $\mu_0 = 4$ and $v = 0.50$ and next 30 observations are generated from the shifted process when

Table 4.3 ARLs of proposed chart when $v = 0.50$, $\mu_0 = 3$, and $ARL_0 = 200$ (Aslam et al., 2016).

c	$k_1 = 3.1489$; $k_2 = 2.0309$; and $i = 2$				$k_1 = 2.8957$; $k_2 = 2.2844$; and $i = 3$			
	$\lambda = 0.05$	$\lambda = 0.20$	$\lambda = 0.50$	$\lambda = 1.00$	$\lambda = 0.05$	$\lambda = 0.20$	$\lambda = 0.50$	$\lambda = 1.00$
0.875	1.13	7.11	51.19	219.90	1.21	8.90	61.46	235.30
0.925	2.94	31.95	124.54	267.38	3.48	37.20	132.34	269.47
0.9375	5.02	49.06	151.39	270.27	6.00	55.06	156.97	270.68
0.95	9.95	75.59	179.04	267.48	11.80	81.48	182.20	266.97
0.9625	22.75	114.17	203.20	258.51	26.05	118.55	204.34	257.69
0.975	57.69	161.74	217.83	243.53	62.08	163.62	217.82	242.84
0.9875	139.54	200.16	217.28	223.51	141.49	200.29	217.03	223.14
1	200.00	200.00	200.00	200.00	200.00	200.00	200.00	200.00
1.0125	112.46	157.76	170.32	174.86	114.18	158.38	170.72	175.20
1.025	41.40	104.62	135.97	149.86	44.02	106.16	136.82	150.48
1.0375	15.72	63.83	103.69	126.37	17.43	65.94	104.95	127.23
1.05	7.00	38.26	76.95	105.29	7.91	40.35	78.50	106.32
1.0625	3.73	23.35	56.41	87.00	4.23	25.10	58.09	88.16
1.075	2.34	14.76	41.32	71.55	2.64	16.11	42.97	72.77
1.125	1.11	3.66	13.19	32.78	1.16	4.08	14.18	33.92

Table 4.4 ARLs of proposed chart when $v = 0.50$, $\mu_0 = 4$, and $ARL_0 = 200$ (Aslam et al., 2016).

c	$k_1 = 2.9123$; $k_2 = 2.1887$; and $i = 2$				$k_1 = 3.3910$; $k_2 = 2.0644$; and $i = 3$			
	$\lambda = 0.05$	$\lambda = 0.20$	$\lambda = 0.50$	$\lambda = 1.00$	$\lambda = 0.05$	$\lambda = 0.20$	$\lambda = 0.50$	$\lambda = 1.00$
0.875	1.01	3.15	23.28	124.94	1.01	2.48	16.83	105.13
0.925	1.63	15.22	73.14	196.77	1.46	11.38	61.92	189.09
0.9375	2.47	25.16	97.21	213.89	2.06	19.49	86.62	209.94
0.95	4.68	42.82	127.36	227.26	3.63	35.17	119.05	226.28
0.9625	11.18	73.79	161.75	234.23	8.54	65.33	157.03	235.04
0.975	32.92	123.82	193.61	232.38	27.03	117.89	192.36	233.73
0.9875	106.31	183.39	210.20	220.62	100.46	182.20	210.57	221.54
1	200.01	200.01	200.01	200.01	200.00	200.00	200.00	200.00
1.0125	87.22	146.03	165.92	173.58	82.07	143.72	164.53	172.51
1.025	24.72	82.94	123.67	145.19	20.61	78.21	120.69	143.12
1.0375	8.34	43.92	86.57	118.08	6.62	39.09	82.45	115.19
1.05	3.67	23.70	59.04	94.24	3.00	20.03	54.57	90.77
1.0625	2.10	13.47	40.17	74.41	1.83	11.02	35.98	70.63
1.075	1.48	8.17	27.64	58.49	1.37	6.60	24.05	54.64
1.125	1.01	2.14	7.69	23.08	1.01	1.87	6.33	20.28

Table 4.5 ARLs of proposed chart when $v = 0.50$, $\mu_0 = 4$, and $ARL_0 = 500$ (Aslam et al., 2016).

| | $k_1 = 3.1136$; $k_2 = 2.5857$; and $i = 2$ | | | | $k_1 = 3.1147$; $k_2 = 2.6403$; and $i = 3$ | | | |
c	$\lambda = 0.05$	$\lambda = 0.20$	$\lambda = 0.50$	$\lambda = 1.00$	$\lambda = 0.05$	$\lambda = 0.20$	$\lambda = 0.50$	$\lambda = 1.00$
0.875	1.04	6.11	62.29	347.31	1.05	5.92	61.25	346.90
0.925	2.34	36.47	186.48	511.83	2.32	35.62	185.33	512.00
0.9375	4.19	61.11	244.73	551.84	4.08	60.05	243.72	552.19
0.95	9.27	104.16	318.01	583.33	8.96	102.95	317.24	583.81
0.9625	24.79	178.93	403.26	599.13	24.10	177.75	402.82	599.69
0.975	76.84	301.58	484.96	592.05	75.71	300.72	484.89	592.56
0.9875	254.41	455.15	528.89	558.17	253.39	454.93	529.04	558.48
1	500.00	500.00	500.00	500.00	500.00	500.00	500.00	500.00
1.0125	202.69	352.20	405.53	426.49	201.54	351.52	405.08	426.13
1.025	53.81	190.52	292.58	349.10	52.81	189.35	291.67	348.40
1.0375	16.54	96.83	197.92	277.15	16.03	95.66	196.76	276.19
1.05	6.36	50.22	130.74	215.73	6.15	49.26	129.54	214.61
1.0625	3.11	27.27	86.35	166.15	3.05	26.57	85.22	164.96
1.075	1.91	15.65	57.71	127.49	1.91	15.16	56.71	126.30
1.125	1.03	3.10	14.06	45.96	1.04	3.04	13.62	45.06

Table 4.6 ARL comparison between the proposed and Saghir and Lin charts when $v = 0.5$ and $ARL_0 = 500$ (Aslam et al., 2016).

| | $\lambda = 0.05$ | | $\lambda = 0.20$ | |
c	Proposed with $k_1 = 3.1136$; $k_2 = 2.5857$; and $i = 2$	Saghir and Lin with $k = 3.0902$	Proposed with $k_1 = 3.1136$; $k_2 = 2.5857$; and $i = 2$	Saghir and Lin with $k = 3.0902$
0.875	1.04	1.13	6.11	9.61
0.925	2.34	3.58	36.47	45.70
0.9375	4.19	6.45	61.11	71.94
0.95	9.27	13.35	104.16	115.91
0.9625	24.79	31.81	178.93	190.01
0.975	76.84	87.12	301.58	309.24

(Continued)

Table 4.6 (Continued)

c	$\lambda = 0.05$		$\lambda = 0.20$	
	Proposed with $k_1 = 3.1136$; $k_2 = 2.5857$; and $i = 2$	Saghir and Lin with $k = 3.0902$	Proposed with $k_1 = 3.1136$; $k_2 = 2.5857$; and $i = 2$	Saghir and Lin with $k = 3.0902$
0.9875	254.41	262.96	455.15	457.01
1	500.00	500.00	500.00	500.00
1.0125	202.69	211.62	352.20	357.38
1.025	53.81	62.00	190.52	199.35
1.0375	16.54	21.41	96.83	105.79
1.05	6.36	8.97	50.22	57.78
1.0625	3.11	4.50	27.27	33.15
1.075	1.91	2.66	15.65	20.05
1.125	1.03	1.08	3.10	4.37

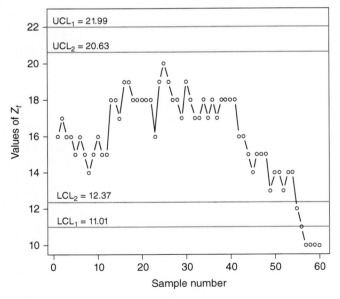

Figure 4.4 Proposed control chart for simulated data (Aslam et al., 2016).

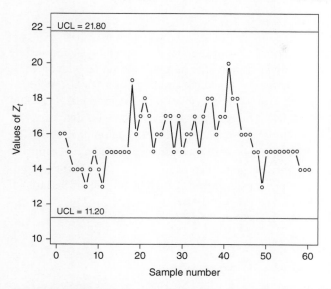

Figure 4.5 Control chart by Saghir and Lin (2014) for simulated data (Aslam et al., 2016).

$c = 0.9375$ in Figure 4.4 (proposed chart) and Figure 4.5 (Saghir & Lin, 2014, chart). By comparing Figures 4.4 and 4.5, it can be noted that the control chart proposed by Aslam et al. (2016) detected shift at the 57th observation, while the control chart proposed by Saghir and Lin (2014) do not detect any shift in the process.

Aslam et al. (2016) also discussed about the application of MDS sampling plan using the COM–Poisson distribution when as $i = 2$, $\lambda = 0.15$, $Z_0 = 19.846$, and $ARL_0 = 500$ using the circuit board data taken from (Montgomery, 2009) having MLE $\hat{\mu} = 2.871$ and $\hat{v} = 0.365$. The control chart for circuit board data shown on Figure 4.6 depicts that although the data came from the in-control process, several observations are close to UCL_2.

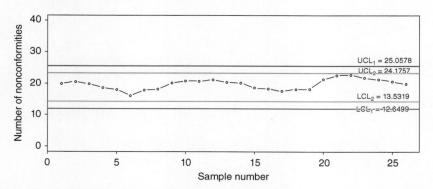

Figure 4.6 Proposed control chart for the circuit board data when $i = 2$, $k_1 = 3.3930$, $k_2 = 2.2247$, $\lambda = 0.15$, and $ARL_0 = 500$ (Aslam et al., 2016).

4.4 Variable Charts

In this section, the variable control chart using the MDS sampling is proposed by assuming that the quality of interest follows the normal distribution with mean, μ, and variance, σ^2. The X-bar chart using the MDS sampling was originally proposed by Aslam, Khan, and Jun (2014) and stated as follows. The operational procedure of X-bar chart using MDS is taken from Aslam et al. (2014) and stated as

Step I: Compete $\overline{X}$ based on n subgroups from the production process.

Step II: If $LCL_2 \leq \overline{X} \leq UCL_2$, declare the process in control and out of control if $\overline{X} \geq UCL_1$ or $\overline{X} \leq LCL_1$. Otherwise, go to Step III.

Step III: Observe whether previous i subgroups are in control state, declare the process is in control and out of control otherwise.

The pair of UCLs and LCLs is given as follows:

$$UCL_1 = m + k_1\sigma/\sqrt{n}, LCL_1 = m - k_1\sigma/\sqrt{n}, \tag{4.34}$$

$$UCL_2 = m + k_2\sigma/\sqrt{n}, LCL_2 = m - k_2\sigma/\sqrt{n} \tag{4.35}$$

Control chart proposed by Aslam et al. (2014) is the extension of the traditional Shewhart control chart and reduces to it when $k_1 = k_2 = k$ and $i = 1$.

The probability of in-control for X-bar chart using the MDS sampling is given by

$$P_{in} = P\left(LCL_2 \leq \overline{X} \leq UCL_2\right) + \left\{P\left(LCL_1 < \overline{X} < LCL_2\right)\right.$$

$$\left. + P\left(UCL_2 < \overline{X} < UCL_1\right)\right\}\left\{P\left(LCL_2 \leq \overline{X} \leq UCL_2\right)\right\}^i \tag{4.36}$$

The simplified form of Eq. (4.36) is given by

$$P_{in} = (2\Phi(k_2) - 1) + 2\{\Phi(k_1) - \Phi(k_2)\}\{(2\Phi(k_2) - 1)\}^i \tag{4.37}$$

The ARL of this chart is given by

$$ARL_0 = \frac{1}{1 - \left\{(2\Phi(k_2) - 1) + 2\{\Phi(k_1) - \Phi(k_2)\}\{(2\Phi(k_2) - 1)\}^i\right\}} \tag{4.38}$$

For the shifted process, m to $m + c\sigma$, we have

$$P_{in}^1 = \left(\Phi\left(k_2 - c\sqrt{n}\right) + \Phi\left(k_2 + c\sqrt{n}\right) - 1\right)$$

$$+ \left\{\left(\Phi\left(-\left(k_2 + c\sqrt{n}\right) - \Phi\left(-\left(k_1 + c\sqrt{n}\right)\right)\right)\right) + \Phi\left(\left(k_1 - c\sqrt{n}\right)\right.\right.$$

$$\left.\left. - \Phi\left(\left(k_2 - c\sqrt{n}\right)\right)\right)\right\}\left\{\Phi\left(k_2 - c\sqrt{n}\right) + \Phi\left(k_2 + c\sqrt{n}\right) - 1\right\}^i \tag{4.39}$$

Table 4.7 ARL_1 for the proposed chart when $r_0 = 370$ (Aslam et al., 2014).

c	$n = 10$ $k_1 = 3.150$ $k_2 = 2.255$ $i = 2$	$n = 20$ $k_1 = 3.282$ $k_2 = 2.250$ $i = 3$	$n = 30$ $k_1 = 3.148$ $k_2 = 2.256$ $i = 2$	$n = 40$ $k_1 = 3.229$ $k_2 = 2.199$ $i = 2$	$n = 50$ $k_1 = 3.131$ $k_2 = 2.274$ $i = 2$
0	370.0	370.0	370.0	370.0	370.0
0.1	230.7	152.8	117.5	87.82	72.42
0.2	90.27	35.89	22.79	13.71	10.36
0.3	34.48	10.34	6.204	3.697	2.855
0.4	14.56	4.115	2.568	1.735	1.465
0.5	7.027	2.224	1.536	1.216	1.118
0.6	3.909	1.531	1.188	1.059	1.025
0.7	2.494	1.237	1.061	1.013	1.003
0.8	1.799	1.101	1.017	1.002	1.000
0.9	1.436	1.039	1.003	1.000	1.000
1	1.238	1.013	1.000	1.000	1.000

The ARL for the chart is reported as

$$ARL_1 = \frac{1}{1 - P_{in}^1} \tag{4.40}$$

The control chart parameters when $r_0 = 370$ are shown in Table 4.7. Table 4.7 shows that there is decreasing trends in ARLs as the values of n increases from 10 to 50. Table 4.8 shows the comparison of Aslam et al. (2014) control chart with the np Shewhart chart. From Table 4.8, it can be seen that the control chart using MDS sampling significantly reduced the values of ARLs.

4.5 Control Charts for Non-normal Distributions

In practice there are many situations in which the form of the distribution is unknown, the data are highly skewed, or the data are collected on individual basis such as health monitoring issues. In such situations, the control charts designed

Table 4.8 Comparisons of ARL_1 for both charts $r_0 = 370$ (Aslam et al., 2014).

c	n = 10		n = 20		n = 30		n = 40		n = 50	
		$k_1 = 3.15$ $k_2 = 2.25$		$k_1 = 3.28$ $k_2 = 2.25$		$k_1 = 3.14$ $k_2 = 2.25$		$k_1 = 3.22$ $k_2 = 2.199$		$k_1 = 3.13$ $k_2 = 2.274$
	$k = 2.99$	$i = 2$	$k = 2.99$	$i = 3$	$k = 2.99$	$i = 2$	$k = 2.99$	$i = 2$	$k = 2.999$	$i = 2$
0	370.0	370.03	370.00	370.01	370.00	370.01	370.00	370.01	370.00	370.032
0.1	243.8	230.75	177.5	152.83	137.01	117.57	109.86	87.822	90.568	72.421
0.2	109.8	90.278	56.547	35.893	35.135	22.795	24.154	13.715	17.719	10.368
0.3	49.57	34.488	20.550	10.348	11.432	6.205	7.398	3.698	5.267	2.856
0.4	24.15	14.561	8.851	4.115	4.777	2.568	3.133	1.735	2.315	1.465
0.5	12.81	7.028	4.493	2.225	2.519	1.536	1.771	1.216	1.421	1.119
0.6	7.398	3.909	2.661	1.531	1.632	1.188	1.271	1.059	1.120	1.025
0.7	4.632	2.494	1.811	1.237	1.253	1.062	1.083	1.013	1.026	1.004
0.8	3.133	1.799	1.392	1.102	1.091	1.017	1.020	1.002	1.004	1.000
0.9	2.278	1.436	1.180	1.039	1.028	1.004	1.004	1.000	1.000	1.000
1	1.771	1.239	1.076	1.013	1.007	1.001	1.000	1.000	1.000	1.000

for the normal distributions cannot be applied for the monitoring of the process. Santiago and Smith (2013) suggested that data having individual observation is highly skewed and modeled using the exponential distribution. In this section, we will discuss some control charts for individual observation using the MDS sampling.

4.6 Control Charts for Exponential Distribution

The time between events data is usually highly skewed. The exponential distribution is a quite suitable distribution to study the time between events (Santiago & Smith, 2013). Santiago and Smith (2013) proposed the control chart for the exponential distortion and named it as T-chart using the single sampling scheme. Aslam, Azam, Khan, and Jun (2015) extended Santiago and Smith (2013) control chart for the MDS sampling.

Suppose that time between events T follows the exponential distribution with scale parameter or mean life θ having following probability density function (p.d.f.)

$$f(t) = \frac{1}{\theta} e^{-\frac{t}{\theta}}, \quad t > 0 \tag{4.41}$$

Johnson and Kotz (1970) suggested that $T^* = T^{1/\beta} \sim WD(\beta, \theta^{1/\beta})$, where T^* is a transformed variable. Nelson (1994) commented that $\beta = 3.6$ is a good normal approximation. Aslam, Azam, et al. (2015) used T^* with $\beta = 3.6$ to design the following T-chart using the MDS sampling.

Step I: Measure the quality of interest T from a selected item from the production process. Compute $T^* = T^{1/3.6}$.

Step II: If $LCL_2 \leq T^* \leq UCL_2$, the process is in control state when $T^* \geq UCL_1$ or $T^* \leq LCL_1$ and study i proceeding subgroups if in control; otherwise, the process is out of control.

The mean and variance of T^* are given by

$$\mu_{T^*} = \theta_0^* \Gamma\left(1 + \frac{1}{3.6}\right) \tag{4.42}$$

$$\sigma_{T^*} = \theta_0^* \sqrt{\Gamma\left(1 + \frac{2}{3.6}\right) - \Gamma\left(1 + \frac{1}{3.6}\right)^2} \tag{4.43}$$

where $\theta_0^* = \theta_0^{1/3.6}$ and $\Gamma(.)$ is a gamma function.

The probability of in-control at θ_0 is given by (Aslam, Azam, et al., 2015)

$$P_{in} = a + (b + c) * a^i \tag{4.44}$$

where $a = P(LCL_2 \leq T^* \leq UCL_2)$, $b = P(LCL_1 < T^* < LCL_2)$, and $c = P(UCL_2 < T^* < UCL_1)$.

The inner and outer control limits are given by

$$\text{LCL}_1 = \theta_0^* c_{\text{L1}} \tag{4.45}$$

$$\text{UCL}_1 = \theta_0^* c_{\text{U1}} \tag{4.46}$$

$$\text{LCL}_2 = \theta_0^* c_{\text{L2}} \tag{4.47}$$

$$\text{UCL}_2 = \theta_0^* c_{\text{U2}} \tag{4.48}$$

where

$$c_{\text{L1}} = \Gamma\left(1 + \frac{1}{3.6}\right) - k_1 \sqrt{\Gamma\left(1 + \frac{2}{3.6}\right) - \Gamma\left(1 + \frac{1}{3.6}\right)^2}$$

$$c_{\text{U1}} = \Gamma\left(1 + \frac{1}{3.6}\right) + k_1 \sqrt{\Gamma\left(1 + \frac{2}{3.6}\right) - \Gamma\left(1 + \frac{1}{3.6}\right)^2}$$

$$c_{\text{L2}} = \Gamma\left(1 + \frac{1}{3.6}\right) - k_2 \sqrt{\Gamma\left(1 + \frac{2}{3.6}\right) - \Gamma\left(1 + \frac{1}{3.6}\right)^2}$$

$$c_{\text{U2}} = \Gamma\left(1 + \frac{1}{3.6}\right) + k_2 \sqrt{\Gamma\left(1 + \frac{2}{3.6}\right) - \Gamma\left(1 + \frac{1}{3.6}\right)^2}$$

where k_1 and k_2 are control chart coefficients.

So for the in-control process, the probabilities are given by

$$a_0 = P(\text{LCL}_2 \leq T^* \leq \text{UCL}_2 \mid \theta = \theta_0) = \exp\left\{-c_{\text{L2}}^\beta\right\} - \exp\left\{-c_{\text{U2}}^\beta\right\} \tag{4.49}$$

$$b_0 = P(\text{LCL}_1 \leq T \leq \text{LCL}_2 \mid \theta = \theta_0) = \exp\left\{-c_{\text{L1}}^\beta\right\} - \exp\left\{-c_{\text{L2}}^\beta\right\} \tag{4.50}$$

$$c_0 = P(\text{UCL}_2 \leq T \leq \text{UCL}_1 \mid \theta = \theta_0) = \exp\left\{-c_{\text{U2}}^\beta\right\} - \exp\left\{-c_{\text{U1}}^\beta\right\} \tag{4.51}$$

The ARL is given by

$$\text{ARL}_0 = \frac{1}{1 - (a_0 + (b_0 + c_0)*a_0^i)} \tag{4.52}$$

The ARL for the process at θ_0 to θ_1 is given by

$$\text{ARL}_1 = \frac{1}{1 - (a_1 + (b_1 + c_1)*a_1^i)} \tag{4.53}$$

where

$$a_1 = \exp\left\{-\frac{\theta_0}{\theta_1} c_{\text{L2}}^\beta\right\} - \exp\left\{-\frac{\theta_0}{\theta_1} c_{\text{U2}}^\beta\right\}$$

$$b_1 = \exp\left\{-\frac{\theta_0}{\theta_1} c_{\text{L1}}^\beta\right\} - \exp\left\{-\frac{\theta_0}{\theta_1} c_{\text{L2}}^\beta\right\}$$

Table 4.9 ARLs of the proposed control chart when $r_0 = 370$ (Aslam, Azam, et al., 2015).

θ_0/θ_1	$i = 1$ $k_1 = 2.761945$ $k_2 = 2.322945$	$i = 2$ $k_1 = 2.8131$ $k_2 = 2.2388$	$i = 3$ $k_1 = 2.8138$ $k_2 = 2.2941$	$i = 4$ $k_1 = 2.8740$ $k_2 = 2.2656$	$i = 5$ $k_1 = 2.8435$ $k_2 = 2.246305$
1	370	370.02	370	370.01	370.02
0.9	231.07	227.89	226.42	224.59	223.59
0.8	133.73	129.48	127.73	125.02	124.17
0.7	73.84	69.91	68.43	65.85	65.38
0.6	39.69	36.66	35.61	33.67	33.46
0.5	21.03	18.97	18.31	17.08	17.01
0.4	11.07	9.81	9.45	8.77	8.78
0.3	5.84	5.15	4.99	4.67	4.71
0.2	3.11	2.8	2.75	2.64	2.68

$$c_1 = \exp\left\{ -\frac{\theta_0}{\theta_1} c_{U2}^\beta \right\} - \exp\left\{ -\frac{\theta_0}{\theta_1} c_{U1}^\beta \right\}$$

Table 4.9 shows the ARLs for various shift ratios θ_0/θ_1 and specified allowed previous subgroup i. From Table 4.9, the decreasing trends in ARLs when i increases can be noted. It means, more the previous subgroups are in-control, quick shift can be expected in the process.

4.7 Control Charts for Gamma Distribution

Suppose now that the time between events T follows the gamma distribution with shape parameter a and scale parameter b with following cumulative distribution function

$$P(T \le t) = 1 - \sum_{j=1}^{a-1} \frac{e^{-\frac{t}{b}}(t/b)^j}{j!} \tag{4.54}$$

Wilson and Hilferty (1931) suggested that $T^* = T^{1/3}$ has approximate normal distribution with mean and variance

$$\mu_{T^*} = \frac{b^{1/3}\Gamma(a + 1/3)}{\Gamma(a)} \tag{4.55}$$

$$\sigma_{T^*} = \frac{b^{2/3}\Gamma(a + 2/3)}{\Gamma(a)} - \mu_{T^*}^2. \tag{4.56}$$

Aslam, Arif, and Jun (2017) proposed the following control chart based on this transformation, which is stated as:

Step I: Measure the quality of interest T from a selected item from the production process. Compute $T^* = T^{1/3}$.

Step II: If $LCL_2 \leq T^* \leq UCL_2$, the process is in control state when $T^* \geq UCL_1$ or $T^* \leq LCL_1$ and study i proceeding subgroups if in control; otherwise, the process is out of control.

The four control limits are given by

$$LCL_1 = b_0^{1/3} LL_1 \tag{4.57}$$

$$UCL_1 = b_0^{1/3} UL_1 \tag{4.58}$$

$$LCL_2 = b_0^{1/3} LL_2 \tag{4.59}$$

$$UCL_2 = b_0^{1/3} UL_2 \tag{4.60}$$

where

$$LL_1 = \left[\frac{\Gamma(a + 1/3)}{\Gamma(a)} - k_1 \sqrt{\frac{\Gamma(a + 2/3)}{\Gamma(a)} - \left(\frac{\Gamma(a + 1/3)}{\Gamma(a)}\right)^2} \right]$$

$$UL_1 = \left[\frac{\Gamma(a + 1/3)}{\Gamma(a)} + k_1 \sqrt{\frac{\Gamma(a + 2/3)}{\Gamma(a)} - \left(\frac{\Gamma(a + 1/3)}{\Gamma(a)}\right)^2} \right]$$

$$LL_2 = \left[\frac{\Gamma(a + 1/3)}{\Gamma(a)} - k_2 \sqrt{\frac{\Gamma(a + 2/3)}{\Gamma(a)} - \left(\frac{\Gamma(a + 1/3)}{\Gamma(a)}\right)^2} \right]$$

$$UL_2 = \left[\frac{\Gamma(a + 1/3)}{\Gamma(a)} + k_2 \sqrt{\frac{\Gamma(a + 2/3)}{\Gamma(a)} - \left(\frac{\Gamma(a + 1/3)}{\Gamma(a)}\right)^2} \right]$$

The probability of in-control at b_0 is given by

$$P_{in}^0 = P(\text{LCL}_2 \leq T^* \leq \text{UCL}_2 \mid b_0) + \{P(\text{LCL}_1 < T^* < \text{LCL}_2 \mid b_0)$$
$$+ P(\text{UCL}_2 < T^* < \text{UCL}_1 \mid b_0)\}\{P(\text{LCL}_2 \leq T^* \leq \text{UCL}_2 \mid b_0)\}^i$$

$$(4.61)$$

here,

$$P(\text{LCL}_2 \leq T^* \leq \text{UCL}_2 \mid b_0) = P(T^* < \text{UCL}_2 \mid b_0) - P(T^* < \text{LCL}_2 \mid b_0)$$

$$P(\text{LCL}_2 \leq T^* \leq \text{UCL}_2 \mid b_0) = \sum_{j=1}^{a-1} \frac{e^{-\text{UL}_2^3}(\text{UL}_2^3)^j}{j!} - \sum_{j=1}^{a-1} \frac{e^{-\text{LL}_2^3}(\text{LL}_2^3)^j}{j!}$$

$$P(\text{LCL}_1 < T^* < \text{LCL}_2 \mid b_0) = \sum_{j=1}^{a-1} \frac{e^{-\text{LL}_2^3}(\text{UL}_2^3)^j}{j!} - \sum_{j=1}^{a-1} \frac{e^{-\text{LL}_1^3}(\text{LL}_1^3)^j}{j!}$$

$$P(\text{UCL}_2 < T^* < \text{UCL}_1 \mid b_0) = \sum_{j=1}^{a-1} \frac{e^{-\text{UL}_1^3}(\text{UL}_1^3)^j}{j!} - \sum_{j=1}^{a-1} \frac{e^{-\text{UL}_2^3}(\text{UL}_2^3)^j}{j!}$$

So the ARL for in-control state is given by

$$\text{ARL}_0 = \frac{1}{1 - \left(P(\text{LCL}_2 \leq T^* \leq \text{UCL}_2 \mid b_0) + \{P(\text{LCL}_1 < T^* < \text{LCL}_2 \mid b_0) + P(\text{UCL}_2 < T^* < \text{UCL}_1 \mid b_0)\}\{P(\text{LCL}_2 \leq T^* \leq \text{UCL}_2 \mid b_0)\}^i\right)}$$

$$(4.62)$$

The probability that the process is in control for the shifted process at $b_1 = cb_0$, where c is constant, is given by

$$P_{in}^1 = P(\text{LCL}_2 \leq T^* \leq \text{UCL}_2 \mid b_1) + \{P(\text{LCL}_1 < T^* < \text{LCL}_2 \mid b_1)$$
$$+ P(\text{UCL}_2 < T^* < \text{UCL}_1 \mid b_1)\}\{P(\text{LCL}_2 \leq T^* \leq \text{UCL}_2 \mid b_1)\}^i \quad (4.63)$$

where

$$P(\text{LCL}_2 \leq T^* \leq \text{UCL}_2 \mid b_1) = P(T^* < \text{UCL}_2 \mid b_1) - P(T^* < \text{LCL}_2 \mid b_1)$$

$$P(\text{LCL}_2 \leq T^* \leq \text{UCL}_2 \mid b_1) = \sum_{j=1}^{a-1} \frac{e^{\frac{-\text{UL}_2^3}{c}}\left(\frac{\text{UL}_2^3}{c}\right)^j}{j!} - \sum_{j=1}^{a-1} \frac{e^{\frac{-\text{LL}_2^3}{c}}\left(\frac{\text{LL}_2^3}{c}\right)^j}{j!}$$

$$P(\text{LCL}_1 < T^* < \text{LCL}_2 \mid b_1) = \sum_{j=1}^{a-1} \frac{e^{\frac{-\text{LL}_2^3}{c}}\left(\frac{\text{UL}_2^3}{c}\right)^j}{j!} - \sum_{j=1}^{a-1} \frac{e^{\frac{-\text{LL}_1^3}{c}}\left(\frac{\text{LL}_1^3}{c}\right)^j}{j!}$$

Table 4.10 ARL$_1$ values for the proposed chart with $i = 2$ when $r_0 = 370$ (Aslam et al., 2017).

c	$a = 2$ $k_1 = 3.470263$ $k_2 = 2.963487$	$a = 5$ $k_1 = 4.621132$ $k_2 = 4.078435$	$a = 10$ $k_1 = 5.862599$ $k_2 = 5.197681$	$a = 20$ $k_1 = 7.385778$ $k_2 = 6.807985$
1.00	370.02	370.01	370.30	370.17
1.01	343.07	332.80	319.30	304.56
1.02	318.56	300.00	276.32	251.92
1.03	296.23	271.03	239.98	209.46
1.04	275.85	245.38	209.14	175.05
1.05	257.23	222.62	182.89	147.04
1.10	184.89	140.90	98.13	66.08
1.15	136.83	93.25	56.67	33.22
1.20	103.89	64.20	34.95	18.47
1.30	63.95	33.69	15.71	7.38
1.40	42.31	19.80	8.47	3.88
1.50	29.68	12.75	5.27	2.49
1.60	21.84	8.84	3.65	1.85
1.70	16.72	6.52	2.76	1.52
1.80	13.23	5.05	2.22	1.33
1.90	10.77	4.07	1.88	1.21
2.00	8.97	3.40	1.66	1.14
2.50	4.64	1.92	1.19	1.02
3.00	3.12	1.46	1.07	1.00

$$P(\text{UCL}_2 < T^* < \text{UCL}_1 \,|\, b_1) = \sum_{j=1}^{a-1} \frac{e^{-\frac{\text{UL}_1^3}{c}}\left(\frac{\text{UL}_1^3}{c}\right)^j}{j!} - \sum_{j=1}^{a-1} \frac{e^{-\frac{\text{UL}_2^3}{c}}\left(\frac{\text{UL}_2^3}{c}\right)^j}{j!}$$

The ARL for the shifted process is given by

$$\text{ARL}_1 = \frac{1}{1 - \left(P(\text{LCL}_2 \leq T^* \leq \text{UCL}_2 | b_1) + \{P(\text{LCL}_1 < T^* < \text{LCL}_2 | b_1) + P(\text{UCL}_2 < T^* < \text{UCL}_1 | b_1)\}\{P(\text{LCL}_2 \leq T^* \leq \text{UCL}_2 | b_1)\}^i\right)}$$

$$(4.64)$$

Tables 4.10 and 4.11 show the values of ARLs when specified ARL is 370; $a = 2$, 5, 10, 20; and $i = 2, 3$. The ARL decreases as i changes from 2 to 3 and a increases from 2 to 20. Aslam et al. (2017) also discussed the advantage of chart with the

Table 4.11 ARL$_1$ values for the proposed chart with $i = 3$ when $r_0 = 370$ (Aslam et al., 2017).

c	$a = 2$ $k_1 = 3.480068$ $k_2 = 2.982044$	$a = 5$ $k_1 = 4.587742$ $k_2 = 4.293158$	$a = 10$ $k_1 = 5.79097$ $k_2 = 5.372559$	$a = 20$ $k_1 = 7.35734$ $k_2 = 6.89495$
1.00	370.12	370.18	370.94	370.66
1.01	342.67	334.91	323.42	306.42
1.02	317.74	303.64	282.86	254.53
1.03	295.05	275.88	248.13	212.43
1.04	274.37	251.15	218.30	178.12
1.05	255.50	229.10	192.60	150.04
1.10	182.42	148.71	107.22	68.01
1.15	134.14	100.61	63.52	34.26
1.20	101.24	70.58	39.81	19.02
1.30	61.66	38.10	18.13	7.57
1.40	40.45	22.79	9.77	3.98
1.50	28.18	14.81	6.03	2.57
1.60	20.64	10.31	4.15	1.92
1.70	15.74	7.60	3.11	1.57
1.80	12.43	5.87	2.48	1.37
1.90	10.11	4.72	2.08	1.25
2.00	8.42	3.92	1.82	1.17
2.50	4.39	2.15	1.26	1.03
3.00	2.99	1.59	1.10	1.00

Shewhart control chart in terms of ARLs given in Table 4.12. Table 4.12 shows that the ARLs from the MDS gamma distribution are smaller than the traditional Shewhart control chart at all values of shape parameter a. The efficiency of the chart for the gamma distribution using MDS is also shown using the simulation study. Figure 4.7 shows the plotting of statistic T^* based on the simulated data having 20 observations generated from the in-control process when $a = 2$ and $b_0 = 1$ and the next 30 observations generated from the shifted process $b_1 = 1.5$. The chart based on the gamma distribution shows the shift in the process at 49th subgroup. Figure 4.8 is presented for the Shewhart control chart for the gamma distribution, which shows no shift in the process.

Table 4.12 Comparison of ARL$_1$ values for proposed chart with $i = 3$ and Shewhart chart when $r_0 = 370$ (Aslam et al., 2017).

	$a = 1$		$a = 2$		$a = 5$		$a = 10$		$a = 20$	
c	Proposed with $k_1 = 2.84$ $k_2 = 2.45$	Shewhart with $k = 2.82$	Proposed with $k_1 = 3.48$ $k_2 = 2.98$	Shewhart with $k = 3.44$	Proposed with $k_1 = 4.58$ $k_2 = 4.29$	Shewhart with $k = 4.57$	Proposed with $k_1 = 5.79$ $k_2 = 5.37$	Shewhart with $k = 5.75$	Proposed with $k_1 = 7.35$ $k_2 = 6.89$	Shewhart with $k = 7.29$
1.00	370.00	370.05	370.12	370.41	370.18	370.06	370.94	370.25	370.66	370.50
1.01	348.35	349.01	342.67	344.82	334.91	335.94	323.42	326.82	306.42	314.37
1.02	328.34	329.54	317.74	321.47	303.64	305.64	282.86	289.40	254.53	268.03
1.03	309.82	311.50	295.05	300.13	275.88	278.68	248.13	257.05	212.43	229.59
1.04	292.65	294.77	274.37	280.60	251.15	254.62	218.30	229.01	178.12	197.55
1.05	276.73	279.23	255.50	262.70	229.10	233.11	192.60	204.61	150.04	170.71
1.10	212.29	216.16	182.42	192.54	148.71	154.23	107.22	121.29	68.01	87.40
1.15	166.51	171.11	134.14	145.21	100.61	106.44	63.52	76.35	34.26	48.93
1.20	133.18	138.11	101.24	112.28	70.58	76.21	39.81	50.61	19.02	29.57
1.30	89.56	94.53	61.66	71.48	38.10	42.82	18.13	25.30	7.57	13.03
1.40	63.63	68.31	40.45	48.74	22.79	26.57	9.77	14.50	3.98	6.99
1.50	47.27	51.54	28.18	35.10	14.81	17.82	6.03	9.25	2.57	4.35
1.60	36.43	40.29	20.64	26.41	10.31	12.72	4.15	6.40	1.92	3.03
1.70	28.93	32.41	15.74	20.60	7.60	9.54	3.11	4.74	1.57	2.30
1.80	23.58	26.72	12.43	16.56	5.87	7.46	2.48	3.69	1.37	1.87
1.90	19.63	22.48	10.11	13.65	4.72	6.03	2.08	3.01	1.25	1.59
2.00	16.66	19.24	8.42	11.48	3.92	5.01	1.82	2.53	1.17	1.41
2.50	8.96	10.65	4.39	6.07	2.15	2.66	1.26	1.51	1.03	1.08
3.00	5.97	7.18	2.99	4.05	1.59	1.87	1.10	1.21	1.00	1.02

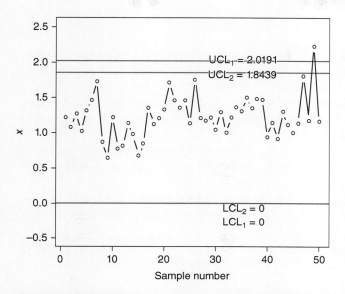

Figure 4.7 Proposed control chart for the simulated data (Aslam et al., 2017).

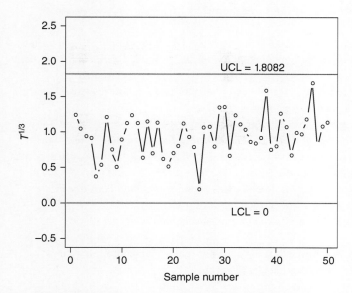

Figure 4.8 Shewhart control chart for the simulated data (Aslam et al., 2017).

References

Aslam, M., Ahmad, L., Jun, C.-H., & Arif, O. H. (2016). A control chart for COM–Poisson distribution using multiple dependent state sampling. *Quality and Reliability Engineering International, 32*(8), 2803–2812.

Aslam, M., Arif, O.-H., & Jun, C.-H. (2017). A control chart for gamma distribution using multiple dependent state sampling. *Industrial Engineering & Management Systems, 16*(1), 109–117.

Aslam, M., Azam, M., Khan, N., & Jun, C.-H. (2015). A control chart for an exponential distribution using multiple dependent state sampling. *Quality & Quantity, 49*(2), 455–462.

Aslam, M., Khan, N., & Jun, C.-H. (2014). A multiple dependent state control chart based on double control limits. *Research Journal of Applied Sciences, Engineering and Technology, 7*(21), 4490–4493.

Aslam, M., Nazir, A., & Jun, C.-H. (2015). A new attribute control chart using multiple dependent state sampling. *Transactions of the Institute of Measurement and Control, 37*(4), 569–576.

Aslam, M., Yen, C.-H., Chang, C.-H., & Jun, C.-H. (2013). Multiple states repetitive group sampling plans with process loss consideration. *Applied Mathematical Modelling, 37*(20), 9063–9075.

Balamurali, S., Jeyadurga, P., & Usha, M. (2017). Optimal designing of multiple deferred state sampling plan for assuring percentile life under Weibull distribution. *The International Journal of Advanced Manufacturing Technology, 93*(9–12), 3095–3109.

Balamurali, S., & Jun, C.-H. (2007). Multiple dependent state sampling plans for lot acceptance based on measurement data. *European Journal of Operational Research, 180*(3), 1221–1230.

Conway, R. W., & Maxwell, W. L. (1962). A queuing model with state dependent service rates. *Journal of Industrial Engineering, 12*(2), 132–136.

del Castillo, J., & Pérez-Casany, M. (2005). Overdispersed and underdispersed Poisson generalizations. *Journal of Statistical Planning and Inference, 134*(2), 486–500.

Famoye, F. (1994). Statistical control charts for shifted generalized Poisson distribution. *Journal of the Italian Statistical Society, 3*(3), 339–354.

Johnson, N. L., & Kotz, S. (1970). *Distributions in statistics: Continuous univariate distributions* (Vol. 2). New York: Hougton Mifflin.

Montgomery, C. D. (2009). *Introduction to statistical quality control* (6th ed.). New York: Wiley.

Nelson, L. S. (1994). A control chart for parts-per-million nonconforming items. *Journal of Quality Technology, 26*(3), 239–240.

Olkin, I. (1992). Generalized Poisson distributions: Properties and applications. *Journal of the American Statistical Association, 87*(420), 1245–1246.

Saghir, A., & Lin, Z. (2014). A flexible and generalized exponentially weighted moving average control chart for count data. *Quality and Reliability Engineering International, 30*(8), 1427–1443.

Santiago, E., & Smith, J. (2013). Control charts based on the exponential distribution: Adapting runs rules for the t chart. *Quality Engineering, 25*(2), 85–96.

Sellers, K. F., Borle, S., & Shmueli, G. (2012). The COM-Poisson model for count data: a survey of methods and applications. *Applied Stochastic Models in Business and Industry, 28*(2), 104–116.

Wilson, E. B., & Hilferty, M. M. (1931). The distribution of chi-square. *Proceedings of the National Academy of Sciences, 17*(12), 684–688.

Wortham, A. W., & Baker, R. C. (1976). Multiple deferred state sampling inspection. *The International Journal of Production Research, 14*(6), 719–731.

5

EWMA Control Charts Using Repetitive Group Sampling Scheme

5.1 Concept of Exponentially Weighted Moving Average (EWMA) Methodology

Roberts (1959) first introduced the EWMA control scheme as an alternative to the Shewhart control chart. Using simulation to evaluate its properties, he showed that the EWMA is useful for detecting small shifts in the mean of a process (Abujiya, Riaz, & Lee, 2013; Chandara, 2001; Hunter, 1986).

EWMA is a weighted average of all past and current observations; it is very insensitive to the assumption of normality (Abbas, 2015). It is, therefore, an ideal replacement for a Shewhart chart when normality cannot be assumed. In the literature of statistical quality control, two schemes have been introduced for the early detection of out-of-control processes: cumulative sum (CUSUM) (Pignatiello & Perry, 2011) and EWMA (Abbasi & Miller, 2013; Allen, 2006). The performance of these schemes in detecting the unusual change in the production process is approximately the same, but the working, operation, and the applicability of the EWMA control chart (Amiri, Moslemi, & Doroudyan, 2015; Zou & Tsung, 2010) are comparatively easier as compared with the CUSUM scheme (Huang, Shu, & Su, 2014; Mavroudis & Nicolas, 2013; Sua, Shub, & Tsui, 2011). Like the CUSUM, EWMA is sensitive to small shifts in the process mean but does not match the ability of a Shewhart chart to detect larger shifts (Haq, Brown, & Moltchanova, 2014; Yang, 2013). For this reason, it is sometimes used together with a Shewhart chart (Akhundjanov & Pascual, 2017; Kruger & Xie, 2012; Montgomery, 2009; Trietsch, 1998; Xie, Goh, & Kuralmani, 2012).

Preamble: The best chart is considered as one that has the efficiency of pointing out the situation of out-of-control process. Two commonly used techniques of quick detection of the out-of-control process have been explained in this chapter.

Introduction to Statistical Process Control, First Edition. Muhammad Aslam, Aamir Saghir, and Liaquat Ahmad.
© 2021 John Wiley & Sons, Inc. Published 2021 by John Wiley & Sons, Inc.

In this section hypothetical data of 20 observations of three different kinds have been collected. The values of data-I are based on descending order, data-II on ascending order, and data-III are randomly selected observations. We will use three different values for λ to determine its effect and to know which value of λ gives changes to the collected data. We know that the value of λ ranges from 0 to 1, i.e. $(0 < \lambda \leq 1)$.

Here we set value of $\lambda = 0.1$, 0.3, and 0.5 to monitor the production process to show step by step the effects of λ on collected data.

The parameter λ determines the rate at which "previous" data enter into the calculation of the EWMA statistic. A value of $\lambda = 1$ implies that only the most recent measurement influences the EWMA. Thus, a large value of λ (closer to 1) gives more weight to recent data and less weight to older data; a small value of λ (closer to 0) gives more weight to older data. By using R-language then graphs have been shown with their control limits.

The weighting parameter λ also called the smoothing parameter, and its value in practice is often selected based on how fast the process mean changes. For more rapid mean changes, one should choose λ to be "large," and for slower (or less frequent) mean changes, one should choose λ to be "small."

The EWMA is defined as

$$w_t = z_t \lambda + (1 - \lambda) w_{t-1}$$

where w_t is the EWMA statistic at time t, z_t is the collected statistic at time t, and λ is the smoothing constant. The upper and lower control limits (UCL, LCL) of the EWMA statistics can be constructed as

$$\text{UCL} = w_t\text{bar} + K * \sigma * \sqrt{\frac{\lambda}{2 - \lambda}}$$

$$\text{LCL} = w_t\text{bar} - K * \sigma * \sqrt{\frac{\lambda}{2 - \lambda}}$$

As mentioned earlier the range of the smoothing constant is 0–1. The pattern of response of λ values can be studied in the following. Let the target value of the mean of z_t values, which is denoted as μ_0, i.e. $\mu_0 = 4.3345$, and the first value of the estimator be 1.2:

Calculation of EWMA statistic with $\lambda = 0.1$:

$$w_t = z_t \lambda + (1 - \lambda) w_{t-1}$$
$$w_1 = 1.2 * 0.1 + (1\text{-}0.1)(4.3345)$$

$$w_1 = 4.02125$$

Using $\sigma = 2.202$ and so on like this, we will get all values for EWMA:

Calculation of control limits for EWMA

$$\text{UCL} = z_t\text{bar} + k * \sigma * \sqrt{\frac{\lambda}{2-\lambda}}$$

$$\text{UCL} = 4.07 + 2.45 * 2.202\sqrt{\frac{0.1}{2-0.1}}$$

$$\text{UCL} = 5.30$$

where z_t bar $= 4.07$ and

$$\text{LCL} = z_t\text{bar} - k * \sigma * \sqrt{\frac{\lambda}{2-\lambda}}$$

$$\text{LCL} = 4.07 - 2.45 * 2.202 * \sqrt{\frac{0.1}{2-0.1}}$$

$$\text{LCL} = 2.832$$

where k is a constant. Here, we fix the value of $k = 2.45$.

Figure 5.1 shows EWMA control chart with UCL $= 5.30$ and LCL $= 2.832$ using $\lambda = 0.1$. It detects 6 out-of-control values in start and last 4 out-of-control values.

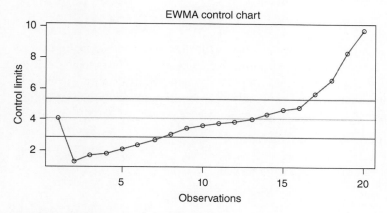

Figure 5.1 EWMA control chart for $\lambda = 0.1$.

It shows if we fix the value of λ closer to 0, then the width of limits will be closer and maximum values will be out of control. So, for this type of data, if the value of λ is larger, so small number of values will be out of control.

Calculation of EWMA statistic with $\lambda = 0.3$:

Let the target value of the mean of z_t values, which is denoted as μ_0, i.e. $\mu_0 = 643$, and the first value of the estimator be 1124:

$$w_t = z_t\,\lambda + (1 - \lambda)\,w_{t-1}$$

$$w_1 = 1124 * 0.3 + (1 - 0.3)\,(643)$$

$$w_1 = 787.3$$

Using $\sigma = 233.186$ and so on like this, we will get all values for EWMA:

Calculation of control limits for EWMA:

$$\text{UCL} = z_t\text{bar} + k * \sigma * \sqrt{\frac{\lambda}{2 - \lambda}}$$

$$\text{UCL} = 654.27 + 2.45 * 233.186\sqrt{\frac{0.3}{2 - 0.3}}$$

$$\text{UCL} = 894.26$$

$$z_t\text{bar} = 654.27$$

$$\text{LCL} = z_t\text{bar} - k * \sigma * \sqrt{\frac{\lambda}{2 - \lambda}}$$

$$\text{LCL} = 654.27 - 2.45 * 233.186\sqrt{\frac{0.3}{2 - 0.3}}$$

$$\text{LCL} = 414.27$$

Figure 5.2 shows EWMA control chart with UCL = 894.26 and LCL = 414.27 using $\lambda = 0.3$. It detects 3 out-of-control values in start and last 4 out-of-control values. It shows if we fix value of λ closer to 0, then the width of limits will be closer and maximum values will be out of control. So, for this type of data, if the value of λ is larger, so small number of values will be out of control.

Calculation of EWMA statistic with $\lambda = 0.5$:

Let the target value of the mean of z_t values, which is denoted as μ_0, i.e. $\mu_0 = 509.70$, and the first value of the estimator be 980:

$$w_t = z_t\,\lambda + (1 - \lambda)\,w_{t-1}$$

$$w_1 = 980 * 0.5 + (1 - 0.5)\,(509.70)$$

$$w_1 = 744.85$$

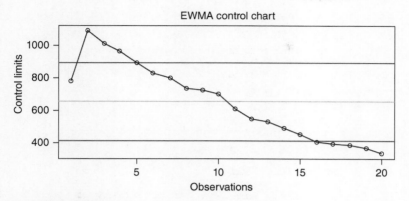

Figure 5.2 EWMA control chart for $\lambda = 0.3$.

Using $\sigma = 199.43$ and so on like this, we will get all values for EWMA:

Calculation of control limits for EWMA:

$$UCL = z_t bar + k * \sigma * \sqrt{\frac{\lambda}{2 - \lambda}}$$

$$UCL = 497.76 + 2.45 * 199.43 \sqrt{\frac{0.5}{2 - 0.5}}$$

$$UCL = 779.87$$

$$z_t bar = 497.76$$

$$LCL = z_t bar - k * \sigma * \sqrt{\frac{\lambda}{2 - \lambda}}$$

$$LCL = 497.76 - 2.45 * 199.43 \sqrt{\frac{0.5}{2 - 0.5}}$$

$$LCL = 215.6$$

Figure 5.3 shows EWMA control chart with UCL = 779.87and LCL = 215.6 using $\lambda = 0.5$. It detects 1 out-of-control value at second observation and overall 3 out-of-control values. It shows if we fix value of λ closer to 0, then the width of limits will be closer and maximum values will be out of control. So, for this type of data, if the value of λ is larger, so small number of values will be out of control.

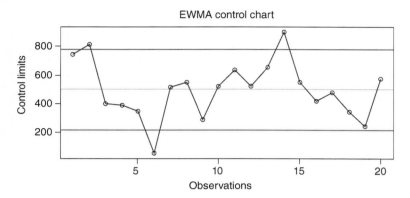

Figure 5.3 EWMA control chart for $\lambda = 0.5$.

5.2 Attraction of EWMA Methodology in Manufacturing Scenario

The control chart methodology is used to detect the off-target items produced by any unit as early as possible to avoid any loss due to rework or the scrap. During recent years the production and manufacturing sector have adopted the fast-speed machinery (Abbas, Riaz, & Does, 2011; S. V. Crowder, 1989; Patel & Divecha, 2011; Tien & Liu, 2011). Thus during a very small interval of time, a huge amount of items have been produced; in any production process if it is facing any deteriorating situation, then the management will have to face not only the financial loss but also the repute and the quality of the product is defamed (Shankar, 1992). Therefore, the quick detection of the shifted process becomes inevitable for the quality control personnel (Knoth & Steinmetz, 2013; Oakland, 2003).

The development of any monitoring scheme is judged by estimating the average run length (ARL) value. ARL may be defined as the average number of samples in the in-control region before it falls or indicates an out-of-control process. The run length is a discrete random variable, distributed to a geometric mass function, and we do not know the time of the specific sample when it falls in the out-of-control region. When the process is in control, a larger value of ARL is estimated and is usually denoted by ARL_0 usually 300, 370, or 500. A smaller value of ARL denoted by ARL_1 is considered as the accepted value of the deteriorated process as it detects the out-of-control process more quickly. The value of ARL_0 is calculated using the following formula:

$$ARL_0 = 1/\text{Probability of out} - \text{of} - \text{control}$$
$$= 1/(1 - \text{Probability of in} - \text{control})$$

If probability of out-of-control process is 0.00333, then $ARL_0 = 300$. It means that the process remains in control until it falsely indicates an out-of-control signal. The second characteristic is calculated when the process is out of control or a shift has already been occurred, and the technique measures the signal of the out-of-control process, which is known as other ARL_1:

$$ARL_1 = 1/\text{Probability of out of control of the shifted process}$$

There are several methods to calculate ARL of the in-control and the out-of-control processes including integral equation method, Markov chain method, Martingale approach, Monte Carlo simulations (MCS), and explicit formulas for EWMA charts, proposed by Areepong (2009), and CUSUM charts, proposed by Busaba, Sukparungsee, and Areepong (2012). The simulation is the most commonly used method that measures the average number of samples needed to reach decision about the manufacturing process that shifted from in-control process to the out-of-control process.

5.3 Development of EWMA Control Chart for Monitoring Averages

In this section a variable control chart using the EWMA technique for the non-normal or the unknown distribution of the underlying process is explained. In such a situation the traditional Shewhart control charts are not suitable since the performance of the proposed chart cannot be developed (J. Y. Liu, Xie, Goh, & Chan, 2007; Mason & Young, 2002; Shamsuzzaman & Wu, 2012).

5.4 Development of EWMA Control Chart for Repetitive Sampling Scheme

EWMA control charts have a wide variety of applications in the control chart literature. EWMA control charts for average and dispersion have a rich literature to understand (Abbas, Riaz, & Does, 2013; Abbasi & Miller, 2013; Su-Fen, Wen-Chi, Tzee-Ming, Chi-Chin, & Smiley, 2011). In addition to this the EWMA scheme has been used in the SS, double sampling, multiple sampling, ranked set sampling, repetitive sampling (Arizono, Okada, Tomohiro, & Takemoto, 2016; Sherman, 1965), bootstrapping sampling, sequential sampling, multiple dependent state sampling, etc. This chapter focuses on the most commonly used sampling scheme of the repetitive sampling scheme, which is considered as the intermediate sampling scheme of the single and the sequential sampling schemes. This sampling scheme is the efficient and acceptable sampling scheme with respect to its application and its understanding.

5.5 EWMA Control Chart for Repetitive Sampling Using Mean Deviation

In this section the dispersion chart using mean absolute deviation from median has been developed for the quick monitoring of the variation in the quality characteristic (Ahmad, Aslam, Arif, & Jun, 2016; Zhang, Li, Chen, & Wang, 2014). We can construct two pairs of control limits with enhanced performance for the quick monitoring (Adeoti & Olaomi, 2018). This performance has been evaluated using the ARLs calculated through simulation when the quality characteristic follows the normal distribution, t-distribution, and the gamma distribution (Kenett, Zacks, & Amberti, 2013). The RGS scheme is quite popular for quick detection of the out-of-control process and is being used very commonly for the last few years (Shankar & Mohapatra, 1993; Soudararajan & Ramasamy, 1984). Many efficient control charts for dispersion have been developed in the literature during last few years, and this is one of the efficient charts using the mean absolute deviation from median. The mean absolute deviation from median using the EWMA technique is better than the range chart and the standard deviation charts because the range chart utilizes only two extreme values, i.e. range $=X_{\mathrm{Max}} - X_{\mathrm{Min}}$, of the data set and neglects the worth of the middle values, thus losing the efficiency, when the sample size is increased. On the other hand, the use of standard deviation as a measure of dispersion has been criticized by many authors as Eddington (1914) and Tukey (1960) because it is very much sensitive to the extreme values in the observations. In addition, the concept of standard deviation is relatively complex (S. Crowder & Hamilton, 1992) and difficult to comprehend as compared with the mean absolute deviation from median, which may be defined as

$$\mathrm{MD} = \frac{\sum\limits_{i=1}^{n} |x_i - \tilde{x}|}{n},$$ where x_i is the quality characteristic of the interest and $\tilde{x}$ is the median of the variable. The equation of the EWMA statistic can be developed as

$W_t = \lambda \mathrm{MD}_t + (1 - \lambda)W_{t-1}$, where λ is defined as the weighting parameter that lies between 0 and 1. The initial value $W_0 = \overline{\mathrm{MD}}$, which is the mean of the mean absolute deviations from median, is thus obtained.

In RGS scheme we have to construct two pairs of control limits known as the outer control limits denoted by UCL_1 and LCL_1 and other one is known as the inner control limits denoted by UCL_2 and LCL_2. When the plotted statistic is located between the inner control limits, the process is declared as the in-control process, and no further action is required for the process. If the plotted statistic is located outside the outer control limits UCL_1 and LCL_1, the process is declared as out-of-control process, and the production is stopped, and the function of rectification is started instantly. If the plotted statistic located inside the outer and inner control limits either between UCL_2 and UCL_1 or between LCL_1 and LCL_2, then decision is waited till the resampling is performed. These limits can be constructed as follows:

$$UCL_1 = \overline{W} + k_1 S_w$$

$$LCL_1 = \overline{W} - k_1 S_w$$

$$UCL_2 = \overline{W} + k_2 S_w$$

$$LCL_2 = \overline{W} - k_2 S_w$$

where $\overline{W}$ is the sample mean, S_w is the standard deviation, and k_1 and k_2 are the control constants.

The changes in W_t statistic are required to be monitored for its sensitivity. The W_t statistic is posted on the control chart with $\overline{W_t}$ as the central line and only two of the control limits UCL_2 and UCL_1 are used for process monitoring. The reason behind these two control limits is based on the concept that only the increase in the dispersion is required to be monitored as the decrease in the dispersion is an indication for the process improvement. Therefore, the area between the central line and UCL_2 of the control chart is considered as the in-control process.

The procedure adopted for the EWMA control chart using RGS scheme for the dispersion can be explained as follows:

Step I: Select a random sample of a suitable size n from the manufacturing process and calculate sample mean absolute deviation from median as

$$MD = \frac{\sum_{i=1}^{n} |x_i - \widetilde{x}|}{n}$$

, where x_i is the quality characteristic of the interest and $\widetilde{x}$ is the median of the variable. Then calculate the EWMA statistic $W_t = \lambda MD_t + (1 - \lambda)W_{t-1}$, where λ is defined as the weighting parameter.

Step II: Declare the process as out of control if the statistic $W_t \geq UCL_1$ or $W_t \leq LCL_1$. Declare the process as in control if the statistic $LCL_2 \leq W_t \leq UCL_2$. Otherwise, move to step I and repeat the process.

The steps for MCS procedure adopted to construct the control limits for the EWMA control chart of dispersion under RGS scheme for estimating ARL may be explained as follows:

1) Calculate the sample mean and variance of the control statistic.
   1.1) Generate a random sample of size n at each subgroup from a selected distribution having the specified parameters for in-control process. Generate 1000 such subgroups. Any of the sample size can be used.

1.2) Calculate W_t for each subgroup, where $W_t = \lambda MD_t + (1 - \lambda)W_{t-1}$.

1.3) Calculate $\overline{W}$ and S_W from 1000 subgroups.

2) Set up control limits.

2.1) Select the initial values of k_1 and k_2, where k_1 and k_2 are the control chart coefficients.

2.2) Generate a random variable at each subgroup from the selected distribution (say, normal distribution) having the specified parameters for in-control process.

2.3) Calculate W_t for tth subgroup.

2.4) Follow the procedure of the proposed control chart and check if the process is declared as out of control. If the process is declared as in control, go to Step 2.5. If the process is declared as out of control, record the number of subgroups so far as the in-control run length.

2.5) Repeat Steps 2.2–2.4 a sufficient number (10,000 say) of times to calculate the in-control ARL. If the in-control ARL is equal to the specified ARL_0, then go to Step 3 with the current values of k_1 and k_2. Otherwise, modify the values of k_1 and k_2 and repeat Steps 2.2 –2.5.

3) Evaluate the out-of-control ARL.

3.1) Generate a random sample of size n for a subgroup from the selected distribution considering a process shift.

3.2) Calculate W_t for tth subgroup.

3.3) Repeat Steps 3.1 and 3.2 until the process is declared as out of control. Record the number of subgroups until this as a run length.

3.4) Repeat the above steps 10,000 times to obtain the ARL.

The number of replications in any simulation is a major challenge for the researcher. There is no hard and fast rule for the selection of the total number of replicates. Many authors have suggested that the minimum number of replicates in the simulation study should not be less than 50,000 replicates, but still many authors have at least 10,000 replicates for the purpose accurate results. In general the size of the replicates depends on the level of shift occurred in any process. So large replicates are required for smaller amount of shift, whereas large shifts can be detected with the smaller number of replicates.

The ARL is used as the performance measure of any developed control chart methodology, which is defined as the expected number of samples before it shows the out-of-control signal. Smaller values are better in detecting the ability of the shifted process, while larger values are desirable for the in-control process as there are least chances of the running process to show out-of-control situation. The measure of ARL of the production process for its performance evaluation has been criticized by many authors including Barnard (1959), Bissell (1969), Gan (1993a, 1993b), and Woodall (1983). They claim that the use of ARL can lead to confusing results if the run length distribution is skewed and non-normal or the shape of the distribution is unknown. In such situation an alternate and more reliable measure, the standard deviation of the run length (SDRL) or the median of the run length distribution (MDRL), can be used. These can very easily be measured whenever the ARL is measured. For the developed control chart of the EWMA using RGS for the dispersion, we have calculated ARL as well as the MDRL and the SDRL.

The ARL of the in-control process can be calculated using the formula

$ARL_0 = \dfrac{1}{\alpha}$, where α is the probability of the type I error.

The ARL of the out-of-control process whenever the shift occurs in the production process can be calculated using the formula

$ARL_1 = \dfrac{1}{1-\beta}$, where β is the probability of the type II error.

The abovementioned methodology has been used for three commonly used probability distributions. Suppose that the quality characteristic under study follows the normal distribution with mean μ and the scale parameter is σ, then, the density function $\phi(x; \mu, \sigma)$ of a normal distribution can be written as

$$N(x; \mu, \sigma) = \phi(x; \mu, \sigma) = \left(1/\sigma\sqrt{2\pi}\right) \exp\left(-(x-\mu)^2/2\sigma^2\right) \qquad -\infty < x < \infty,$$

In real-life situation, the assumptions of the normally distributed quality characteristic are either not known or never met, so two other distributions, i.e. the heavy tailed symmetrical Student's t-distribution and the positively skewed gamma distribution, are also considered. Their probability distributions are given as

$$t(x; \nu) = \phi(x; \nu) = \frac{\Gamma\left(\dfrac{\nu+1}{2}\right)}{\sqrt{\nu * \pi}\,\Gamma(\nu/2)} \left(1 + \frac{x^2}{\nu}\right)^{-\left(\frac{\nu+1}{2}\right)} \qquad -\infty < x < \infty, \nu > 0,$$

$$\text{Gamma}(x; \alpha, \beta) = \Phi(x; \alpha, \beta) = \frac{\beta^\alpha}{\Gamma(\alpha)} x^{\alpha-1} e^{-\beta x}, \qquad x > 0$$

Using R-code program the run length characteristics have been calculated for $n = 5$, $\lambda = 0.05$ and $ARL_0 = 200$ for these three distributions and shown in Tables 5.1–5.6. The graphical comparison of the proposed and existing schemes have been shown in Figures 5.4–5.8.

Table 5.1 RL characteristics of the EWMA-MD chart under repetitive sampling for normally distributed quality characteristics when $\lambda = 0.05$ and $ARL_0 = 200$ (Ahmad et al., 2016).

Shift (δ)		$n = 5$ $k_1 = 2.325424,$ $k_2 = 1.481746$	$n = 10$ $k_1 = 3.73735,$ $k_2 = 1.98823$	$n = 15$ $k_1 = 3.259351,$ $k_2 = 2.064653$
1.00	ARL_1	200.00	200.00	200.00
	MDRL	39.00	22.00	21.00
	SDRL	62.92	48.94	39.17
1.01	ARL_1	47.79	25.74	16.04
	MDRL	26.00	11.00	10.00
	SDRL	55.18	35.32	33.43
1.02	ARL_1	35.68	15.64	13.28
	MDRL	18.00	5.00	4.00
	SDRL	44.46	25.20	24.11
1.03	ARL_1	28.53	10.69	9.32
	MDRL	11.00	2.00	2.00
	SDRL	39.74	20.34	18.53
1.04	ARL_1	21.49	6.54	5.41
	MDRL	6.50	1.00	1.00
	SDRL	33.11	14.66	12.95
1.05	ARL_1	14.64	3.04	2.26
	MDRL	4.00	1.00	1.00
	SDRL	23.83	8.76	6.17
1.06	ARL_1	10.69	1.80	1.31
	MDRL	2.00	1.00	1.00
	SDRL	20.82	5.31	2.98
1.07	ARL_1	8.06	1.17	1.02
	MDRL	1.00	1.00	1.00
	SDRL	17.30	1.95	0.42
1.08	ARL_1	4.81	1.07	1.00
	MDRL	1.00	1.00	1.00
	SDRL	12.08	1.10	0.06
1.09	ARL_1	3.43	1.00	1.00
	MDRL	1.00	1.00	1.00
	SDRL	9.35	0.00	0.00
1.10	ARL_1	2.36	1.00	1.00
	MDRL	1.00	1.00	1.00
	SDRL	6.65	0.00	0.00

Table 5.2 RL characteristics of the EWMA-MD chart under repetitive sampling for normally distributed quality characteristics for $n = 5$ and $ARL_0 = 200$ (Ahmad et al., 2016).

Shift (δ)		$\lambda = 0.25$ $k_1 = 2.250371,$ $k_2 = 2.127114$	$\lambda = 0.50$ $k_1 = 3.010836,$ $k_2 = 2.84386$	$\lambda = 0.75$ $k_1 = 3.505799,$ $k_2 = 2.853409$
1.00	ARL_1	200.00	200.00	200.00
	MDRL	71.00	72.00	85.00
	SDRL	83.36	85.85	93.28
1.10	ARL_1	26.15	32.26	42.85
	MDRL	17.00	23.50	32.00
	SDRL	25.89	31.57	39.66
1.20	ARL_1	8.95	14.12	19.81
	MDRL	6.00	10.00	14.00
	SDRL	9.31	12.94	19.07
1.30	ARL_1	4.03	7.78	10.79
	MDRL	2.00	6.00	7.00
	SDRL	4.88	7.33	10.48
1.40	ARL_1	2.00	4.85	6.81
	MDRL	1.00	3.00	5.00
	SDRL	2.39	5.01	6.37
1.50	ARL_1	1.21	3.25	4.63
	MDRL	1.00	2.00	3.00
	SDRL	0.96	3.23	4.17
1.60	ARL_1	1.04	2.34	3.43
	MDRL	1.00	1.00	2.00
	SDRL	0.36	2.34	3.03
1.70	ARL_1	1.00	1.73	2.90
	MDRL	1.00	1.00	2.00
	SDRL	0.06	1.54	2.67
1.80	ARL_1	1.00	1.50	2.30
	MDRL	1.00	1.00	2.00
	SDRL	0.00	1.16	1.85
1.90	ARL_1	1.00	1.32	1.95
	MDRL	1.00	1.00	1.00
	SDRL	0.00	0.83	1.54
2.00	ARL_1	1.00	1.19	1.77
	MDRL	1.00	1.00	1.00
	SDRL	0.00	0.61	1.28

Table 5.3 RL characteristics of the EWMA-MD chart under repetitive sampling for t-distributed quality characteristics when $\lambda = 0.05$ and $ARL_0 = 200$ (Ahmad et al., 2016).

Shift (δ)		$n = 5$ $k_1 = 2.248201,$ $k_2 = 1.468618$	$n = 10$ $k_1 = 2.234471,$ $k_2 = 1.366732$	$n = 15$ $k_1 = 2.633052,$ $k_2 = 1.514487$
1.00	ARL_1	201.00	201.00	200.00
	MDRL	38.00	13.00	17.00
	SDRL	61.29	36.09	48.65
1.01	ARL_1	43.23	18.32	19.69
	MDRL	25.00	5.00	7.00
	SDRL	49.78	28.96	28.33
1.02	ARL_1	32.31	11.27	9.94
	MDRL	14.00	2.00	2.00
	SDRL	41.57	20.68	18.64
1.03	ARL_1	23.93	5.79	5.71
	MDRL	10.00	1.00	1.00
	SDRL	33.59	12.68	13.40
1.04	ARL_1	19.96	3.53	2.44
	MDRL	6.00	1.00	1.00
	SDRL	33.22	9.08	7.13
1.05	ARL_1	12.06	1.72	1.30
	MDRL	3.00	1.00	1.00
	SDRL	20.41	6.13	3.09
1.06	ARL_1	9.97	1.19	1.12
	MDRL	2.00	1.00	1.00
	SDRL	18.90	2.14	1.52
1.07	ARL_1	6.50	1.02	1.02
	MDRL	1.00	1.00	1.00
	SDRL	14.80	0.25	0.73
1.08	ARL_1	4.01	1.00	1.00
	MDRL	1.00	1.00	1.00
	SDRL	9.25	0.03	0.00
1.09	ARL_1	2.63	1.00	1.00
	MDRL	1.00	1.00	1.00
	SDRL	6.83	0.00	0.00
1.10	ARL_1	1.86	1.00	1.00
	MDRL	1.00	1.00	1.00
	SDRL	4.46	0.00	0.00

Table 5.4 RL characteristics of the EWMA-MD chart under repetitive sampling for t-distributed quality characteristics for $n = 5$ and $ARL_0 = 200$ (Ahmad et al., 2016).

Shift (δ)		$\lambda = 0.25$ $k_1 = 3.334963,$ $k_2 = 2.464435$	$\lambda = 0.50$ $k_1 = 3.655528,$ $k_2 = 2.599459$	$\lambda = 0.75$ $k_1 = 3.745567,$ $k_2 = 2.712584$
1.00	ARL_1	201.00	200.00	201.00
	MDRL	63.00	67.50	85.00
	SDRL	75.74	79.73	89.24
1.1	ARL_1	23.44	30.23	40.21
	MDRL	16.00	20.00	28.00
	SDRL	23.33	30.70	40.66
1.2	ARL_1	8.93	13.45	18.75
	MDRL	5.00	10.00	13.00
	SDRL	9.73	12.38	18.18
1.3	ARL_1	3.83	7.26	9.82
	MDRL	2.00	5.00	7.00
	SDRL	4.73	7.32	9.67
1.4	ARL_1	1.85	4.40	6.69
	MDRL	1.00	3.00	5.00
	SDRL	2.23	4.38	6.30
1.5	ARL_1	1.22	3.08	4.55
	MDRL	1.00	2.00	3.00
	SDRL	0.99	3.08	3.98
1.6	ARL_1	1.03	2.22	3.52
	MDRL	1.00	1.00	2.00
	SDRL	0.28	2.04	3.09
1.7	ARL_1	1.00	1.77	2.76
	MDRL	1.00	1.00	2.00
	SDRL	0.00	1.65	2.29
1.8	ARL_1	1.00	1.44	2.27
	MDRL	1.00	1.00	2.00
	SDRL	0.00	1.02	1.88
1.9	ARL_1	1.00	1.25	1.94
	MDRL	1.00	1.00	1.00
	SDRL	0.00	0.75	1.43
2.0	ARL_1	1.00	1.17	1.73
	MDRL	1.00	1.00	1.00
	SDRL	0.00	0.64	1.21

Table 5.5 RL characteristics of the EWMA-MD chart under repetitive sampling for Gamma distributed quality characteristics when $\lambda = 0.05$ and $ARL_0 = 200$ (Ahmad et al., 2016).

Shift (δ)		$n = 5$ $k_1 = 2.58425,$ $k_2 = 2.482192$	$n = 10$ $k_1 = 2.881411,$ $k_2 = 2.550056$	$n = 15$ $k_1 = 2.940985,$ $k_2 = 1.847446$
1.00	ARL_1	201.00	200.00	201.00
	MDRL	158.02	145.40	133.53
	SDRL	100.10	75.74	53.47
1.01	ARL_1	48.76	22.38	8.95
	MDRL	55.50	24.00	11.00
	SDRL	107.32	75.32	46.22
1.02	ARL_1	45.01	20.45	9.42
	MDRL	44.00	21.00	10.00
	SDRL	103.70	71.87	48.12
1.03	ARL_1	43.33	14.52	7.37
	MDRL	42.50	13.00	6.00
	SDRL	101.94	76.77	41.17
1.04	ARL_1	39.42	11.90	6.61
	MDRL	31.00	10.00	7.00
	SDRL	97.62	66.50	39.46
1.05	ARL_1	35.05	10.39	5.86
	MDRL	30.00	9.00	6.00
	SDRL	92.26	66.64	49.22
1.06	ARL_1	30.81	8.52	7.23
	MDRL	26.00	6.00	5.00
	SDRL	94.23	71.63	44.50
1.07	ARL_1	27.74	6.67	5.58
	MDRL	21.00	3.00	1.00
	SDRL	85.44	55.40	42.49
1.08	ARL_1	23.83	5.45	4.02
	MDRL	11.50	1.00	1.00
	SDRL	25.93	6.27	7.34
1.09	ARL_1	20.88	4.69	3.02
	MDRL	1.00	1.00	1.00
	SDRL	14.00	0.00	0.00
1.10	ARL_1	18.88	3.27	1.32
	MDRL	1.00	1.00	1.00
	SDRL	8.82	0.00	0.00

Table 5.6 RL characteristics comparison of the proposed vs. the existing EWMA-MD chart under repetitive sampling for Gamma distributed quality characteristics when $\lambda = 0.05$ and $ARL_0 = 200$ (Ahmad et al., 2016).

Shift (δ)		n = 5		n = 10		n = 15	
		Existing	Proposed	Existing	Proposed	Existing	Proposed
1.0	ARL_1	200.01	201.00	199.07	200.00	199.71	201.00
	MDRL	138.00	158.02	138.00	145.40	142.00	133.53
	SDRL	206.32	100.10	196.97	75.74	192.92	53.47
1.1	ARL_1	44.74	18.88	28.58	3.27	22.40	1.32
	MDRL	34.00	1.00	23.00	1.00	18.00	1.00
	SDRL	38.41	8.82	21.33	0.00	15.63	0.00

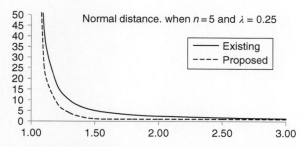

Figure 5.4 ARL_1 for the proposed and existing chart for normal distance for $n = 5$ and $\lambda = 0.25$.

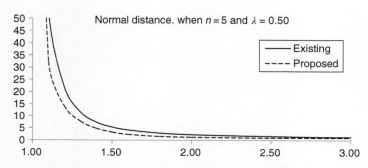

Figure 5.5 ARL_1 for the proposed and existing chart for normal distance for $n = 5$ and $\lambda = 0.50$.

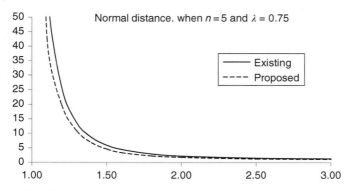

Figure 5.6 ARL$_1$ for the proposed and existing chart for normal distance for n = 5 and λ = 0.75.

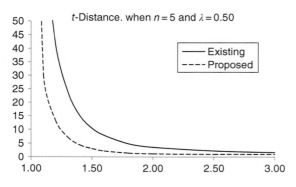

Figure 5.7 ARL$_1$ for the proposed and existing chart for t-distance for n = 5 and λ = 0.50.

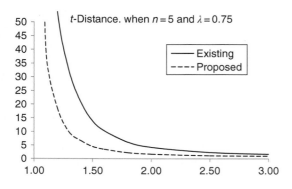

Figure 5.8 ARL$_1$ for the proposed and existing chart for t-distance for n = 5 and λ = 0.75.

5.6 EWMA Control Chart for Sign Statistic Using the Repetitive Sampling Scheme

In this section we presented a nonparametric sign control chart using EWMA statistic for the RGS scheme (Aslam, Azam, & Jun, 2014). In practice there are many situations when the observations under study do not follow the normal distribution or the form of the distribution is skewed or the form of the distribution is unknown (Abdu & Ramli, 2011; Fallah Nezhad, 2012; Graham, Chakraborti, & Human, 2011; Teh, Khoo, & Wu, 2011). It is common practice in control chart literature that we assume that the quality characteristic follows the normal distribution (Fang, Wang, & Deng, 2013; Gan, 1990; Ou, Wu, Khoo, & Chen, 2015). When this assumption is not fully met, the results inferred from the proposed control chart may lead to erroneous conclusions (Abbasi, 2012; Aslam, Azam, & Jun, 2014; Kenett et al., 2013; Mukherjee, Chakraborti, & Graham, 2012; Park & Jun, 2015). In such scenario we have to rely on the commonly alternate technique of the nonparametric distributions. The control chart technique is developed for effective monitoring of the production process when the quality characteristic is distribution-free (Khaliq, Riaz, & Ahmad, 2016; Lovegrove, Valencia, Treasure, Sherlaw-Johnson, & Gallivan, 1997).

Construction of EWMA Sign Chart

Let X be the observations of the interested quality characteristics and T be the target value. Suppose $Y = (X - T)$ is defined as the deviation of the process from the target value T. Further we define the process proportion as $p = P(Y > 0)$. The process is regarded to be in control when $p = 0.5$, and the process is regarded as out of control if $p \neq 0.5 = p_1$. For the purpose of monitoring the process, we select a random sample $(X_1, X_2, X_3, ..., X_n)$ of size n at each subgroup from the process. Let we define

$$I_j = \begin{cases} 1, & \text{if } Y_j = X_j - T > 0 \\ 0, & \text{otherwise} \end{cases} \qquad \text{for } j = 1, 2, 3, ..., n$$

Let the total number of values for $Y_j > 0$ be denoted by M, then

$$M = \sum_{j=0}^{n} I_j$$

Obviously here M follows the binomial probability distribution with the parameters n and $p = 0.5$ for the in-control process. Here M is the sequentially recorded value of the process, and the EWMA sign statistic is calculated as

$$\text{EWMA}_{M_i} = \lambda_{M_i} + (1 - \lambda)\text{EWMA}_{M_{i-1}}$$

where λ is the weighting constant that lies between 0 and 1.

Then the mean and variance of thus defined statistic can be written as

$$\text{Mean} = E(\text{EWMA}_{M_i}) = {}^{n}\!/\!_{2} \text{ and}$$

$$\text{Variance} = \text{Var}(\text{EWMA}_{M_i}) = \frac{\lambda}{2 - \lambda}\left({}^{n}\!/\!_{4}\right)$$

The repetitive sampling of the EWMA sign chart can be explained in the following steps:

Step I: Select a random sample of size n from the production for the ith subgroup and calculate the following EWMA sign statistic as

$$\text{EWMA}_{M_i} = \lambda_{M_i} + (1 - \lambda)\text{EWMA}_{M_{i-1}},$$

Step II: Declare the process as in control if $\text{LCL}_2 \leq \text{EWMA}_{M_i} \leq \text{UCL}_2$. Declare the process as out of control if $\text{EWMA}_{M_i} \geq \text{UCL}_1$ or $\text{EWMA}_{M_i} \leq \text{LCL}_1$. Otherwise go to Step I and repeat the process of sampling.

In repetitive sampling the operational procedure of the EWMA sign chart is built on the following four limits as LCL_1, LCL_2, UCL_2, and UCL_1. Thus two outer and inner control limits can be described as

$$\text{UCL}_1 = \frac{n}{2} + k_1\sqrt{\frac{\lambda}{2 - \lambda}\left(\frac{n}{4}\right)}$$

$$\text{LCL}_1 = \frac{n}{2} - k_1\sqrt{\frac{\lambda}{2 - \lambda}\left(\frac{n}{4}\right)}$$

$$\text{UCL}_2 = \frac{n}{2} + k_2\sqrt{\frac{\lambda}{2 - \lambda}\left(\frac{n}{4}\right)}$$

$$\text{LCL}_2 = \frac{n}{2} - k_2\sqrt{\frac{\lambda}{2 - \lambda}\left(\frac{n}{4}\right)}$$

where k_1 and k_2 are the control coefficients to be determined with $k_1 \geq k_2$. It is to be emphasized that if $k_1 = k_2$ then the developed chart is transformed to be the SS scheme chart.

The MCS is used to determine the control coefficients of the EWMA sign chart for RGS scheme because the theoretical derivation of the ARLs is not possible directly. As mentioned earlier ARL values are computed for any developed control chart to evaluate its performance. When the ARL values of the shifted process (ARL$_1$) are smaller as compared with another chart values, then it is known as the better one in quick detection of the out-of-control process. Here ARL$_1$ values are computed for various combinations of n and λ against the specified in-control ARL values (ARL$_0$) for $p = 0.5$.

The complete MCS procedure can be listed as

1) Set up control limits.
2) Select the sample size (or subgroup size) n and the smoothing constant λ.
3) Select the initial values of k_1 and k_2.
4) Generate a random variable M_i for ith subgroup from the binomial distribution having parameters n and $p = 0.5$.
5) Calculate EWMA$_{M_i}$ for ith subgroup.
6) Follow the procedure of the described control chart and check if the process is declared as out of control. If the process is declared as in control, go to Step 4. If the process is declared as out of control, record the number of subgroups so far as the in-control run length.
7) Repeat Steps 4–6 a sufficient number (10,000 say) of times to calculate the in-control ARL. If the in-control ARL is equal to the specified ARL$_0$, then go to Step 2 with the current values of k_1 and k_2. Otherwise, modify the values of k_1 and k_2 and repeat Steps 4–7 for the shifted process.
8) Evaluate the out-of-control ARL.
9) Generate a random variable M_i for ith subgroup from the binomial distribution with parameters n and $p = p_1 \neq 0.5$ considering a shift.
10) Calculate EWMA$_{M_i}$ for ith subgroup.
11) Repeat Steps 9 and 10 until the process is declared as out of control. Record the number of subgroups as an out-of-control run length.
12) Repeat the above steps a sufficient number (10,000 say) of times to obtain the out-of-control ARL.

Using the abovementioned steps in R-code program (Kenett et al., 2013), the ARL values are computed as given in Tables 5.7 and 5.8.

The comparison of the EWMA sign chart for RGS scheme with the SS scheme has been made, and the ARL are given in Table 5.7.

The developed EWMA sign chart for RGS has been studied using an example of soft drink taken from Montgomery (2009) and given in Table 5.9.

Table 5.7 Estimation of control chart coefficients K_1 and K_2.

The k_1 and k_2 values of the proposed chart for $ARL_0 \approx 370$

λ

n	0.05		0.10		0.15		0.20		0.25		0.30		0.40		0.50	
	k_1	k_2	k_1	k_2	k_1	k_2	k_1	k_2	k_1	k_2	k_1	k_2	k_1	k_2	k_1	k_2
9	2.09	0.78	2.19	0.84	2.21	0.84	2.49	0.71	2.31	0.77	2.08	0.74	2.10	0.87	2.36	0.89
10	2.03	0.82	2.47	0.89	2.27	0.71	2.06	0.79	2.19	0.84	2.28	0.71	2.35	0.77	2.30	0.72
11	2.09	0.81	2.22	0.81	2.07	0.87	2.06	0.84	2.47	0.73	2.31	0.76	2.20	0.79	2.34	0.94
12	2.13	0.82	2.06	0.87	2.16	0.84	2.27	0.77	2.21	0.74	2.10	0.77	2.02	0.79	2.03	0.95
13	2.09	0.71	2.03	0.78	2.29	0.79	2.27	0.73	2.17	0.79	2.22	0.90	2.01	0.89	2.47	0.99
14	2.16	0.71	2.33	0.84	2.02	0.79	2.03	0.76	2.29	0.84	2.16	0.84	2.06	0.82	2.09	0.70
15	2.27	0.74	2.19	0.71	2.28	0.71	2.21	0.71	2.09	0.78	2.16	0.84	2.06	0.81	2.07	0.86
16	2.34	0.83	2.06	0.75	2.37	0.83	2.16	0.84	2.02	0.82	2.05	0.77	2.15	0.86	2.11	0.82
17	2.13	0.72	2.02	0.88	2.12	0.80	2.35	0.71	2.14	0.71	2.50	0.72	2.16	0.74	2.59	0.97
18	2.10	0.84	2.17	0.73	2.18	0.75	2.24	0.84	2.11	0.73	2.18	0.79	2.19	0.88	2.01	0.75
19	2.49	0.77	2.20	0.85	2.25	0.70	2.12	0.78	2.01	0.79	2.31	0.92	2.03	0.99	2.47	0.73
20	2.08	0.89	2.19	0.74	2.07	0.90	2.04	0.89	2.37	0.83	2.28	0.83	2.73	0.71	2.03	1.00
21	2.04	0.73	2.01	0.72	2.20	0.71	2.45	0.72	2.04	0.80	2.01	0.79	2.34	0.96	2.01	0.72
22	2.16	0.84	2.01	0.70	2.12	0.78	2.13	0.77	2.15	0.89	2.13	0.82	2.39	0.82	2.02	0.83
23	2.02	0.81	2.20	0.80	2.26	0.78	2.15	0.81	2.22	0.71	2.05	0.75	2.21	0.76	2.01	0.85
24	2.04	0.88	2.32	0.74	2.39	0.81	2.02	0.84	2.07	0.79	2.06	0.90	2.42	0.73	2.15	0.97
25	2.19	0.87	2.15	0.78	2.24	0.75	2.23	0.73	2.31	0.76	2.19	0.83	2.02	0.71	2.17	0.74

ARL$_1$ for the proposed chart when $\lambda = 0.05$

n										P_1									
	0.05	0.10	0.15	0.20	0.25	0.30	0.35	0.40	0.45	0.50	0.55	0.60	0.65	0.70	0.75	0.80	0.85	0.90	0.95
9	3	3	4	4	5	7	9	15	36	370	36	15	9	7	5	4	4	3	3
10	3	3	4	4	5	6	9	14	33	370	33	14	9	6	5	4	4	3	3
11	2	3	3	4	5	6	8	13	32	370	32	13	8	6	5	4	3	3	3
12	2	3	3	4	5	6	8	13	30	370	30	13	8	6	5	4	3	3	3
13	2	3	3	4	4	6	8	12	28	370	29	12	8	6	4	4	3	3	2
14	2	3	3	4	4	5	7	12	27	370	27	12	7	5	4	4	3	3	2
15	2	3	3	3	4	5	7	11	26	370	26	11	7	5	4	3	3	3	2
16	2	3	3	3	4	5	7	11	25	370	25	11	7	5	4	3	3	3	2
17	2	2	3	3	4	5	7	11	24	370	24	11	7	5	4	3	3	2	2
18	2	2	3	3	4	5	6	10	23	370	23	10	6	5	4	3	3	2	2
19	2	2	3	3	4	5	6	10	22	370	22	10	6	5	4	3	3	2	2
20	2	2	3	3	4	5	6	10	22	370	22	10	6	5	4	3	3	2	2
21	2	2	3	3	4	4	6	9	21	370	21	9	6	4	4	3	3	2	2
22	2	2	3	3	3	4	6	9	20	370	20	9	6	4	3	3	3	2	2
23	2	2	2	3	3	4	6	9	20	370	20	9	6	4	3	3	2	2	2
24	2	2	2	3	3	4	6	9	19	370	19	9	5	4	3	3	2	2	2
25	2	2	2	3	3	4	5	8	19	370	19	8	5	4	3	3	2	2	2

Source: Reproduced with permission of Elsevier.

Table 5.8 ARL values of the single sampling scheme.

	Comparison of proposed chart with EWMA sign chart in Yang (2013) when $\lambda = 0.05$																			
	P_1																			
	Proposed control chart										EWMA sign chart in Yang et al. [22]									
n	0.05	0.10	0.15	0.20	0.25	0.30	0.35	0.40	0.45	0.50	0.50	0.45	0.40	0.35	0.30	0.25	0.20	0.15	0.10	0.05
9	3	3	4	4	5	7	9	15	36	370	384	57	21	12	9	7	5	5	4	4
10	3	3	4	4	5	6	9	14	33	370	371	52	19	11	8	6	5	4	4	3
11	2	3	3	4	5	6	8	13	32	370	370	48	18	11	8	6	5	4	4	3
12	2	3	3	4	5	6	8	13	30	370	380	45	17	10	7	6	5	4	4	3
13	2	3	3	4	4	6	8	12	28	370	377	43	16	10	7	6	5	4	3	3
14	2	3	3	4	4	5	7	12	27	370	378	40	15	9	7	5	4	4	3	3
15	2	3	3	3	4	5	7	11	26	370	386	39	15	9	7	5	4	4	3	3
16	2	3	3	3	4	5	7	11	25	370	371	36	14	9	6	5	4	3	3	3
17	2	2	3	3	4	5	7	11	24	370	384	35	14	8	6	5	4	3	3	3
18	2	2	3	3	4	5	6	10	23	370	375	34	13	8	6	5	4	3	3	3
19	2	2	3	3	4	5	6	10	22	370	388	33	13	8	6	5	4	3	3	3
20	2	2	3	3	4	5	6	10	22	370	389	32	12	8	6	4	4	3	3	3
21	2	2	3	3	4	4	6	9	21	370	379	30	12	7	5	4	4	3	3	2
22	2	2	3	3	3	4	6	9	20	370	383	29	12	7	5	4	4	3	3	2
23	2	2	2	3	3	4	6	9	20	370	383	28	11	7	5	4	4	3	3	2
24	2	2	2	3	3	4	6	9	19	370	381	27	11	7	5	4	4	3	3	2
25	2	2	2	3	3	4	5	8	19	370	377	27	11	7	5	4	4	3	3	2

Table 5.9 Example of soft drink (Montgomery, 2009).

Subgroup	X_1	X_2	X_3	X_4	X_5	X_6	X_7	X_8	X_9	X_{10}	M_i	EWMA$_{M_i}$
1	0.25	0.5	2	-1	1	-1	0.5	1.5	0.5	-1.5	7	5.10
2	0	0	0.5	1	1.5	1	-1	1	1.5	-1	6	5.15
3	1.5	1	–	-1	0	-1.5	-1	-1	1	-1	4	5.09
4	0	0.5	-2	0	-1	1.5	-1.5	0	-2	-1.5	2	4.93
5	0	0	0	-0.5	0.5	1	-0.5	-0.5	0	0	2	4.79
6	1	-0.5	0	0	0	0.5	-1	1	-2	1	4	4.75
7	1	-1	-1	-1	0	1.5	0	1	0	0	3	4.66
8	0	-1.5	-0.5	1.5	0	0	0	-1	0.5	-0.5	2	4.53
9	-2	-15	1.5	1.5	0	0	0.5	1	0	1	5	4.55
10	-0.5	3.5	0	-1	-1.5	-1.5	-1	-1	1	0.5	3	4.47
11	0	1.5	0	0	2	-1.5	0.5	-0.5	2	-1	4	4.45
12	0	-2	-0.5	0	-0.5	2	1.5	0	0.5	-1	3	4.38
13	-1	-0.5	-0.5	-1	0	0.5	0.5	-1.5	-1	-1	2	4.26
14	0.5	1	-1	-0.5	-2	-1	-1.5	0	1.5	1.5	4	4.25
15	1	0	1.5	1.5	1	-1	0	1	-2	-1.5	5	4.24

The EWMA sign chart for RGS has been shown for $n = 10$ in Figure 5.9. The EWMA sign chart for SS has been shown for $n = 10$ in Figure 5.10.

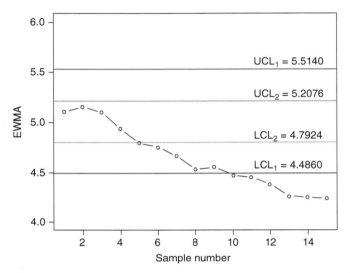

Figure 5.9 EWMA sign chart for RGS for $n = 10$. *Source:* Reproduced with permission of Elsevier.

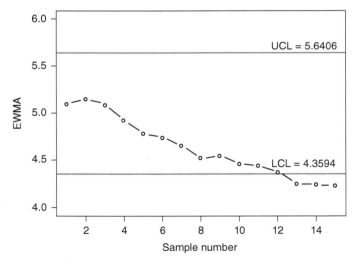

Figure 5.10 EWMA sign chart for SS for $n = 10$. *Source:* Reproduced with permission of Elsevier.

5.7   Designing of a Hybrid EWMA (HEWMA) Control Chart Using Repetitive Sampling

In this section we presented a HEWMA control chart for the RGS scheme (Azam, Aslam, & Jun, 2015). In general the control chart is developed either for large process shifts as the traditional Shewhart control chart or the small process shifts as the EWMA or CUSUM charts (Daryabari, Hashemian, Keyvandarian, & Maryam, 2017; Haq, 2013; Kazemzadeh, Karbasian, & Babakhani, 2013). When both the larger and smaller process shifts are targeted jointly, then hybrid control chart is used (Abbas, Riaz, & Ronald, 2014; Jun, Lee, Lee, & Balamurali, 2010). In this section a HEWMA control chart is explained using two separate EWMA charts.

Construction of HEWMA Control Chart
Let X_1, X_2, X_3, ... be independently and identically distributed random variable from the normal distribution with mean μ and variance σ^2. For two sequences $\{E_1, E_2, E_3, ...\}$ and $\{H_1, H_2, H_3, ...\}$ for two smoothing constants $\lambda_1 \in [0, 1]$ and $\lambda_2 \in [0, 1]$, we define

$$E_t = \lambda_2 X_t + (1 - \lambda_2)E_{t-1} \quad \text{and}$$
$$HE_t = \lambda_1 E_t + (1 - \lambda_1)HE_{t-1}$$

where E_t is any EWMA statistic and HE_t is another EWMA statistic created by E_t. Therefore, the latter type of EWMA statistic is known as HEWMA. Using the RGS scheme (Radhakrishnan & Sivakumaran, 2008) we can construct two pairs of control limits in the following steps:

Step I: Take a random sample of size 1 from the current subgroup at time t and measure the interested quality characteristic X_t. Then calculate

$$HE_t = \lambda_1 E_t + (1 - \lambda_1)HE_{t-1}$$

Step II: Declare the process as in control if $LCL_2 \leq HE_t \leq UCL_2$ (LCL_2 and UCL_2) are called the inner control limits. Declare the process as out of control if $HE_t \geq UCL_1$ or $HE_t \leq LCL_1$ (LCL_1 and UCL_1) are called the outer control limits. Otherwise go to Step I and repeat the process of sampling.

There are three directions for a single sample. As mentioned in Step II, the process may be in control, the process may be out of control, or we have to repeat the process of selecting the sample if we are undecided or if $LCL_1 \leq HE_t \leq LCL_2$ or $UCL_2 \leq HE_t \leq UCL_1$. Therefore, the repetitive sampling scheme provides us a decision with smaller error (S.-W. Liu & Wu, 2014). The frequency of sampling may not be an issue in this computerized era as most of production processes are in surveillance through automatic computer system.

Let the mean of the process be set to $\mu = \mu_0$ when the process is in control. Then the mean and variance of HE_t can be written as

$$E(HE_t) = \mu_0$$

and

$$\mathrm{Var}(HE_t) = \frac{\lambda_1^2 \lambda_2 \sigma^2}{2 - \lambda_2} \left[\frac{1 - (1 - \lambda_1)^{2t}}{\lambda_1(2 - \lambda_1)} - \frac{(1 - \lambda_2)^2 \left\{ (1 - \lambda_2)^{2t} - (1 - \lambda_1)^{2t} \right\}}{(1 - \lambda_2)^2 - (1 - \lambda_1)^2} \right]$$

where t is larger value and then the variance is reduced to

$$\mathrm{Var}(HE_t) = \frac{\lambda_1 \lambda_2 \sigma^2}{(2 - \lambda_2)(2 - \lambda_1)}$$

This formula of variance will be used in the development of the four control limits to be measured in the HEWMA repetitive sampling control chart as

$$\mathrm{UCL}_1 = \mu_0 + k_1 \sigma \sqrt{\frac{\lambda_1 \lambda_2}{(2 - \lambda_2)(2 - \lambda_1)}}$$

$$\mathrm{LCL}_1 = \mu_0 - k_1 \sigma \sqrt{\frac{\lambda_1 \lambda_2}{(2 - \lambda_2)(2 - \lambda_1)}}$$

$$\mathrm{UCL}_2 = \mu_0 + k_2 \sigma \sqrt{\frac{\lambda_1 \lambda_2}{(2 - \lambda_2)(2 - \lambda_1)}}$$

$$\mathrm{LCL}_2 = \mu_0 - k_2 \sigma \sqrt{\frac{\lambda_1 \lambda_2}{(2 - \lambda_2)(2 - \lambda_1)}}$$

where k_1 and k_2 are the control constants to be computed.

It is worth to mention here that if $k_1 = k_2$, then the HEWMA repetitive sampling chart reduces to the HEWMA SS chart.

If $P^0_{\mathrm{out},1}$ denotes the probability that the process is declared as out of control when actually the process is in control, then the following relation can be developed:

$$P^0_{\mathrm{out},1} = P(HE_t < \mathrm{LCL}_1 \mid \mu_0) + P(HE_t > \mathrm{UCL}_1 \mid \mu_0)$$

$$P^0_{\mathrm{out},1} = 2(1 - \Phi(k_1))$$

Here $\Phi(\cdot)$ is the cumulative distribution function of a standard normal distribution. The probability of in-control process for repetition is given as

$$P^0_{\mathrm{rep}} = P\left(\mathrm{UCL}_2 < HE_t < \mathrm{UCL}_1 \mid \mu_0\right) + P(\mathrm{LCL}_1 < HE_t < \mathrm{LCL}_2 \mid \mu_0)$$

Now using the cumulative distribution function of a standard normal distribution, the above equation can be written as

$$P^0_{\mathrm{rep}} = \Phi(k_1) - \Phi(k_2) + \Phi(-k_2) - \Phi(-k_1)$$

$$= 2(\Phi(k_1) - \Phi(k_2))$$

Then the probability that the process is out of control when actually the process is in control is expressed as

$$P_{out}^0 = \frac{P_{out,1}^0}{1 - P_{rep}^0}$$

$$= \frac{2(1 - \Phi(k_1))}{1 - 2(\Phi(k_1) - \Phi(k_2))}$$

For the purpose of calculating the performance of the developed chart, we have to calculate the ARL of the chart. The ARL for the in-control process may be calculated as

$$\text{ARL}_0 = \frac{1}{P_{out}^0}$$

In RGS scheme the decision about the in-control or the out-of-control process is made on basis of the multiple samples, so to calculate the average number of samples required to reach a decision, we must calculate the average number of samples used to reach the decision. The ASS of the in-control process denoted by ASS_0 is calculated as

$$\text{ASS}_0 = \frac{1}{1 - P_{rep}^0}$$

When a process is adjusted to the state of in-control process, it has no surety of running in the form of in-control process for long because there are so many uncontrolled factors that may influence the process. Now, suppose that the mean of the in-control process μ_0 has been shifted to

$$\mu_1 = \mu_0 - c * \sigma,$$

where μ_1 denotes the mean of the shifted process, c is the shift size, and σ is the process standard deviation. Here we suppose that $\mu_0 = 0$ and $\sigma = 1$. The probability that the process is declared as out-of-control process when shift of size c occurs by using the single sample denoted by $P_{out,1}^1$ is given as

$$P_{out,1}^1 = P(HE_t < \text{UCL}_1 \mid \mu_1) + P(HE_t > \text{UCL}_1 \mid \mu_1)$$

After little simplification the above equation can be written as

$$P_{out,1}^1 = \Phi\left(-k_1 + \frac{c}{\sqrt{\frac{\lambda_1 \lambda_2}{(2 - \lambda_2)(2 - \lambda_1)}}} \right) + 1 - \Phi\left(k_1 + \frac{c}{\sqrt{\frac{\lambda_1 \lambda_2}{(2 - \lambda_2)(2 - \lambda_1)}}} \right)$$

Likewise the probability of repetition of the shifted process is given as

$$P_{\text{rep}}^1 = P(\text{UCL}_2 < HE_t < \text{UCL}_1 \mid \mu_1) + P(\text{LCL}_1 < HE_t < \text{LCL}_2 \mid \mu_1)$$

The above equation is simplified as

$$P_{\text{rep}}^1 = \Phi\left(\frac{k_1\sqrt{\frac{\lambda_1\lambda_2}{(2-\lambda_2)(2-\lambda_1)}} + c}{\sqrt{\frac{\lambda_1\lambda_2}{(2-\lambda_2)(2-\lambda_1)}}}\right) - \Phi\left(\frac{k_2\sqrt{\frac{\lambda_1\lambda_2}{(2-\lambda_2)(2-\lambda_1)}} + c}{\sqrt{\frac{\lambda_1\lambda_2}{(2-\lambda_2)(2-\lambda_1)}}}\right)$$
$$+ \Phi\left(\frac{-k_2\sqrt{\frac{\lambda_1\lambda_2}{(2-\lambda_2)(2-\lambda_1)}} + c}{\sqrt{\frac{\lambda_1\lambda_2}{(2-\lambda_2)(2-\lambda_1)}}}\right) - \Phi\left(\frac{-k_1\sqrt{\frac{\lambda_1\lambda_2}{(2-\lambda_2)(2-\lambda_1)}} + c}{\sqrt{\frac{\lambda_1\lambda_2}{(2-\lambda_2)(2-\lambda_1)}}}\right)$$

Then the probability that the shifted process is declared as out of control is given as

$$P_{\text{out}}^1 = \frac{P_{\text{out},1}^1}{1 - P_{\text{rep}}^1}$$

Hence the ARL of the shifted process for being out of control is given as

$$\text{ARL}_1 = \frac{1}{P_{\text{out}}^1}$$

Likewise the ASS of the shifted process denoted by ASS_1 is given as

$$\text{ASS}_1 = \frac{1}{1 - P_{\text{rep}}^1}$$

Now we have to calculate the two control chart constants k_1 and k_2 by using the nonlinear optimization problem. In general the chart constants are calculated by considering the type I error. In this case we determine the two control chart constants k_1 and k_2 so as to minimize the ASS when the ARL of the in-control process should be larger than the targeted value. So the following nonlinear optimization problem is developed to determine the two control chart constants as

$$\text{Minimize ASS}_0 = \frac{1}{1 - P_{\text{rep}}^0} \text{ subject to}$$

$$\text{ARL}_0 = \frac{1}{P_{\text{out}}^1} \geq r_0$$

Here r_0 is the predefined ARL value of the in-control process.

Using the abovementioned equations for the specified smoothing constants λ_1 and λ_2 with the various combination of possible values, the ARL_1 and ASS_1 values are given in Tables 5.10–5.14 for different shifts (Figures 5.11 and 5.12).

Table 5.10 ARL$_1$ of the proposed chart when $\lambda_1 = 0.10$ and $\lambda_2 = 0.25$.

Shift c	$r_0 = 200, k_1 = 2.9265,$ $k_2 = 0.9999$		$r_0 = 300, k_1 = 3.0523,$ $k_2 = 0.9999$		$r_0 = 370, k_1 = 3.1147,$ $k_2 = 0.9998$	
	ASS$_1$	ARL$_1$	ASS$_1$	ARL$_1$	ASS$_1$	ARL$_1$
0.00	1.46	200.16	1.46	301.65	1.46	371.75
0.01	1.46	187.48	1.47	281.16	1.47	345.62
0.02	1.48	156.55	1.49	231.83	1.49	283.20
0.03	1.52	120.72	1.52	175.93	1.52	213.20
0.04	1.56	88.93	1.57	127.46	1.57	153.20
0.05	1.62	64.05	1.63	90.32	1.63	107.69
0.06	1.70	45.72	1.71	63.44	1.71	75.06
0.07	1.79	32.59	1.81	44.50	1.81	52.24
0.08	1.90	23.28	1.92	31.28	1.93	36.43
0.09	2.03	16.73	2.06	22.09	2.07	25.52
0.10	2.17	12.11	2.21	15.71	2.23	17.99
0.15	2.89	2.99	3.10	3.46	3.19	3.76
0.20	2.76	1.35	3.46	1.42	3.28	1.45

Source: Reproduced with permission of Springer Nature.

Table 5.11 ARL$_1$ of the proposed chart when $\lambda_1 = 0.10$ and $\lambda_2 = 0.50$.

Shift c	$r_0 = 200, k_1 = 2.9265,$ $k_2 = 0.9999$		$r_0 = 300, k_1 = 3.0526,$ $k_2 = 0.9999$		$r_0 = 370, k_1 = 3.1142,$ $k_2 = 0.9999$	
	ASS$_1$	ARL$_1$	ASS$_1$	ARL$_1$	ASS$_1$	ARL$_1$
0.00	1.46	200.17	1.46	301.88	1.46	370.95
0.01	1.46	194.56	1.46	292.79	1.46	359.39
0.02	1.47	179.24	1.47	268.11	1.47	328.12
0.03	1.48	157.83	1.49	234.02	1.49	285.15
0.04	1.50	134.24	1.51	197.01	1.51	238.84
0.05	1.53	111.42	1.53	161.75	1.53	195.06
0.06	1.56	91.00	1.56	130.67	1.57	156.75
0.07	1.60	73.58	1.60	104.52	1.60	124.73
0.08	1.64	59.16	1.65	83.15	1.65	98.72
0.09	1.69	47.43	1.70	65.96	1.70	77.93
0.10	1.75	37.99	1.76	52.29	1.77	61.46
0.15	2.14	12.82	2.18	16.64	2.20	19.12
0.20	2.65	4.82	2.77	5.85	2.83	6.50

Source: Reproduced with permission of Springer Nature.

Table 5.12 ARL$_1$ of the proposed chart when $\lambda_1 = 0.10$ and $\lambda_2 = 0.25$.

	$r_0 = 200, k_1 = 2.9267, k_2 = 0.9999$		$r_0 = 300, k_1 = 3.0524, k_2 = 0.9999$		$r_0 = 370, k_1 = 3.1136, k_2 = 0.9999$	
Shift c	ASS$_1$	ARL$_1$	ASS$_1$	ARL$_1$	ASS$_1$	ARL$_1$
0.00	1.46	200.08	1.46	300.16	1.46	371.46
0.01	1.47	168.61	1.48	249.85	1.48	307.22
0.02	1.53	110.65	1.53	159.81	1.53	193.94
0.03	1.62	65.27	1.63	91.77	1.63	109.86
0.04	1.75	37.42	1.76	51.26	1.77	60.56
0.05	1.93	21.50	1.95	28.68	1.96	33.44
0.06	2.15	12.56	2.19	16.28	2.21	18.71
0.07	2.40	7.53	2.48	9.46	2.51	10.70
0.08	2.66	4.70	2.79	5.70	2.85	6.34
0.09	2.88	3.10	3.07	3.62	3.17	3.94
0.10	2.99	2.20	3.26	2.46	3.40	2.62
0.15	1.99	1.07	2.21	1.08	2.34	1.08
0.20	1.23	1.00	1.29	1.00	1.32	1.00

Source: Reproduced with permission of Springer Nature.

Table 5.13 ARL comparison of proposed control chart with Haq (2013) when $\lambda_1 = 0.10$ and $r_0 = 370$.

	$\lambda_2 = 0.50$		$\lambda_2 = 0.25$		$\lambda_2 = 0.10$	
Shift c	Haq	Proposed	Haq	Proposed	Haq	Proposed
0.00	370.35	370.95	371.57	371.75	370.68	371.46
0.01	360.21	359.39	348.57	345.62	313.86	307.22
0.02	332.58	328.12	292.77	283.20	210.28	193.94
0.03	294.14	285.15	228.62	213.20	129.48	109.86
0.04	252.00	238.84	171.94	153.20	79.22	60.56
0.05	211.39	195.06	127.52	107.69	49.48	33.44
0.06	175.12	156.75	94.50	75.06	31.80	18.71
0.07	144.16	124.73	70.43	52.24	21.08	10.70
0.08	118.42	98.72	52.96	36.43	14.41	6.34
0.09	97.33	77.93	40.24	25.52	10.15	3.94
0.10	80.18	61.46	30.92	17.99	7.37	2.62

Source: Reproduced with permission of Springer Nature.

Table 5.14 Simulated data for the application.

Before shift				After shift			
X_t	HE_t	X_t	HE_t	X_t	HE_t	X_t	HE_t
0.8627	0.0086	0.0089	−0.0230	0.3329	0.0470	−1.2804	0.0877
−0.9631	0.0059	0.9766	−0.0171	−1.6110	0.0289	1.0198	0.1081
−0.1377	0.0023	0.5906	−0.0063	0.6676	0.0206	0.0726	0.1243
0.6019	0.0053	0.1684	0.0042	0.5377	0.0190	−0.2211	0.1339
0.3727	0.0114	0.2260	0.0149	0.5507	0.0231	1.8343	0.1588
0.1744	0.0180	−0.3968	0.0195	−0.7274	0.0189	−0.1808	0.1755
−0.9792	0.0134	0.4695	0.0277	1.1066	0.0264	−0.2372	0.1849
−0.7014	0.0025	0.8341	0.0424	−0.2913	0.0292	−0.7842	0.1828
−0.4785	−0.0112	0.0935	0.0548	0.9048	0.0403	1.4765	0.1941
0.4287	−0.0178	−1.5382	0.0489	2.5073	0.0739	2.1682	0.2229

Source: Reproduced with permission of Springer Nature.

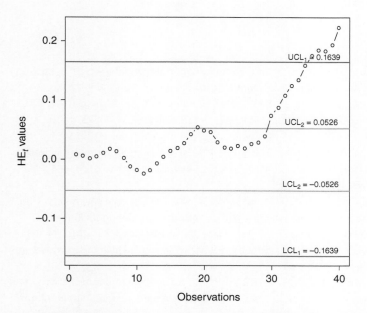

Figure 5.11 HEWMA control chart using repetitive sampling for simulated data.

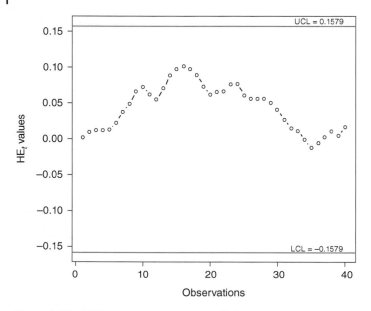

Figure 5.12 HEWMA control chart using SS for simulated data.

References

Abbas, N. (2015). Progressive mean as a special case of exponentially weighted moving average. *Quality and Reliability Engineering International, 31*(4), 719–720.

Abbas, N., Riaz, M., & Does, R. J. M. M. (2011). Enhancing the performance of EWMA charts. *Quality and Reliability Engineering International, 27*(6), 821–833. doi:10.1002/qre.1175

Abbas, N., Riaz, M., & Does, R. J. M. M. (2013). CS-EWMA chart for monitoring process dispersion. *Quality and Reliability Engineering International, 29*(5), 653–663. doi:10.1002/qre.1414

Abbas, N., Riaz, M., & Ronald, J. M. M. D. (2014). An EWMA-type control chart for monitoring the process mean using auxiliary information. *Communications in Statistics – Theory and Methods, 43*(16), 3485. doi:10.1080/03610926.2012.700368

Abbasi, S. A. (2012). A new nonparametric EWMA sign control chart. *Expert Systems with Applications, 39*, 8503.

Abbasi, S. A., & Miller, A. (2013). MDEWMA chart: An efficient and robust alternative to monitor process dispersion. *Journal of Statistical Computation and Simulation, 83*(2), 247–268.

Abdu, M. A. A., & Ramli, R. (2011). A comparison study on the performances of X, EWMA and CUSUM control charts for skewed distributions using weighted standard deviations methods. *Journal of Science and Technology, 3*(2), 77–88.

Abujiya, M. R., Riaz, M., & Lee, M. H. (2013). Enhancing the performance of combined Shewhart-EWMA charts. *Quality and Reliability Engineering International, 29*(8), 1093–1106. doi:10.1002/qre.1461

Adeoti, O. A., & Olaomi, J. O. (2018). Capability index based control chart for monitoring process mean using repetitive sampling. *Communications in Statistics – Theory and Methods, 47*(2), 493–507.

Ahmad, L., Aslam, M., Arif, O., & Jun, C.-H. (2016). Dispersion chart for some popular distributions under repetitive sampling. *Journal of Advanced Mechanical Design, Systems, and Manufacturing, 10*(4), 1–18.

Akhundjanov, S. B., & Pascual, F. G. (2017). Exponentially weighted moving average charts for correlated multivariate Poisson processes. *Communications in Statistics – Theory and Methods, 46*(10), 4977–5000. doi:10.1080/03610926.2015.1096392

Allen, T. T. (2006). *Introduction to engineering statistics and six sigma: statistical quality control and design of experiments and systems*. London: Springer-Verlag London Limited.

Amiri, A., Moslemi, A., & Doroudyan, M. H. (2015). Robust economic and economic-statistical design of EWMA control chart. *The International Journal of Advanced Manufacturing Technology, 78*(1), 511–523. doi:10.1007/s00170-014-6667-9

Areepong, Y. (2009). An integral equation approach for analysis of control charts (PhD). University of Technology, Sydney.

Arizono, I., Okada, Y., Tomohiro, R., & Takemoto, Y. (2016). Rectifying inspection for acceptable quality loss limit based on variable repetitive group sampling plan. *The International Journal of Advanced Manufacturing Technology, 85*(9), 2413–2423. doi:10.1007/s00170-015-8090-2

Aslam, M., Azam, M., & Jun, C. H. (2014). A new exponentially weighted moving average sign chart using repetitive sampling. *Journal of Process Control, 24*(7), 1149–1153. doi:10.1016/j.jprocont.2014.05.001

Azam, M., Aslam, M., & Jun, C. H. (2015). Designing of a hybrid exponentially weighted moving average control chart using repetitive sampling. *The International Journal of Advanced Manufacturing Technology, 77*(9-12), 1927–1933.

Barnard, G. A. (1959). Control charts and stochastic processes. *Journal of the Royal Statistical Society. Series B (Methodological)*, 239–271.

Bissell, A. (1969). Cusum techniques for quality control. *Journal of the Royal Statistical Society: Series C (Applied Statistics), 18*(1), 1–25.

Busaba, J., Sukparungsee, S., & Areepong, Y. (2012). Numerical approximations of average run length for AR (1) on exponential CUSUM. *Computer Science and Telecommunications, 19*, 23.

Chandara, M. J. (2001). *Statistical quality control*. Boca Raton, FL: CRC Press LLC.

Crowder, S., & Hamilton, M. (1992). An EWMA for monitoring a process standard deviation. *Journal of Quality Technology, 24*(1), 12–21.

Crowder, S. V. (1989). Design of exponentially weighted moving average schemes. *Journal of Quality Technology, 21*(3), 155–162.

Daryabari, S. A., Hashemian, S. M., Keyvandarian, A., & Maryam, S. A. (2017). The effects of measurement error on the MAX EWMAMS control chart. *Communications in Statistics – Theory and Methods, 46*(12), 5766–5778. doi:10.1080/03610926.2015.1112911

Eddington, A. S. (1914). *Stellar movements and the structure of the universe.* London: Macmillan.

Fallah Nezhad, M. (2012). A new EWMA monitoring design for multivariate quality control problem. *The International Journal of Advanced Manufacturing Technology, 62*(5–8), 751–758. doi:10.1007/s00170-011-3821-5

Fang, J., Wang, H., & Deng, W. (2013). Design of EWMA control charts for assuring predetermined production process quality. *Research Journal of Applied Sciences, Engineering and Technology, 5*(10), 3010–3014.

Gan, F. (1990). Monitoring observations generated from a binomial distribution using modified exponentially weighted moving average control chart. *Journal of Statistical Computation and Simulation, 37*(1–2), 45–60.

Gan, F. (1993a). An optimal design of EWMA control charts based on median run length. *Journal of Statistical Computation and Simulation, 45*(3-4), 169–184.

Gan, F. (1993b). An optimal design of CUSUM control charts for binomial counts. *Journal of Applied Statistics, 20*(4), 445–460.

Graham, M. A., Chakraborti, S., & Human, S. W. (2011). A nonparametric EWMA sign chart for location based on individual measurements. *Quality Engineering, 23*(3), 227–241. doi:10.1080/08982112.2011.575745

Haq, A. (2013). A new hybrid exponentially weighted moving average control chart for monitoring process mean. *Quality and Reliability Engineering International, 29*(7), 1015–1025. doi:10.1002/qre.1453

Haq, A., Brown, J., & Moltchanova, E. (2014). Improved fast initial response features for exponentially weighted moving average and cumulative sum control charts. *Quality and Reliability Engineering International, 30*(5), 697–710. doi:10.1002/qre.1521

Huang, W. P., Shu, L. J., & Su, Y. (2014). An accurate evaluation of adaptive exponentially weighted moving average schemes. *IIE Transactions, 46*(5), 457–469. doi:10.1080/0740817x.2013.803642

Hunter, J. S. (1986). The exponentially weighted moving average. *Journal of Quality Technology, 18*(4), 203–210.

Jun, C.-H., Lee, H., Lee, S.-H., & Balamurali, S. (2010). A variables repetitive group sampling plan under failure-censored reliability tests for Weibull distribution. *Journal of Applied Statistics, 37*(3), 453–460. doi:10.1080/02664760802715914

Kazemzadeh, R. B., Karbasian, M., & Babakhani, M. A. (2013). An EWMA *t* chart with variable sampling intervals for monitoring the process mean. *The International Journal of Advanced Manufacturing Technology, 66*(1), 125–139. doi:10.1007/s00170-012-4311-0

Kenett, R., Zacks, S., & Amberti, D. (2013). *Modern industrial statistics: With applications in R, MINITAB and JMP*. Hoboken, NJ: Wiley.

Khaliq, Q.-U.-A., Riaz, M., & Ahmad, S. (2016). On designing a new Tukey-EWMA control chart for process monitoring. *The International Journal of Advanced Manufacturing Technology, 82*(1), 1–23. doi:10.1007/s00170-015-7289-6

Knoth, S., & Steinmetz, S. (2013). EWMA *p* charts under sampling by variables. *International Journal of Production Research, 51*(13), 3795–3807. doi:10.1080/00207543.2012.746799

Kruger, U., & Xie, L. (2012). *Advances in statistical monitoring of complex multivariate processes: With applications in industrial process control*. Hoboken, NJ: Wiley.

Liu, J. Y., Xie, M., Goh, T. N., & Chan, L. Y. (2007). A study of EWMA chart with transformed exponential data. *International Journal of Production Research, 45*(3), 743–763. doi:10.1080/00207540600792598

Liu, S.-W., & Wu, C.-W. (2014). Design and construction of a variables repetitive group sampling plan for unilateral specification limit. *Communications in Statistics – Simulation and Computation, 43*(8), 1866–1878.

Lovegrove, J., Valencia, O., Treasure, T., Sherlaw-Johnson, C., & Gallivan, S. (1997). Monitoring the results of cardiac surgery by variable life-adjusted display. *The Lancet, 350*(9085), 1128–1130.

Mason, R. L., & Young, J. C. (2002). *Multivariate statistical process control with industrial applications* (Vol. 9). Philadelphia, USA: Siam.

Mavroudis, E., & Nicolas, F. (2013). EWMA control charts for monitoring high yield processes. *Communications in Statistics – Theory and Methods, 42*(20), 3639. doi:10.1080/03610926.2011.635256

Montgomery, C. D. (2009). *Introduction to statistical quality control* (6th ed.). New York, NY: Wiley.

Mukherjee, A., Chakraborti, S., & Graham, M. A. (2012). Distribution-free exponentially weighted moving average control charts for monitoring unknown location. *Computational Statistics & Data Analysis, 56*(8), 2539–2561.

Oakland, J. S. (2003). *Statistical process control* (6th ed.). Oxford, UK: Butterworth-Heinemann.

Ou, Y., Wu, Z., Khoo, M. B. C., & Chen, N. (2015). A rational sequential probability ratio test control chart for monitoring process shifts in mean and variance. *Journal of Statistical Computation and Simulation, 85*(9), 1765–1781. doi:10.1080/00949655.2014.901327

Park, J., & Jun, C.-H. (2015). A new multivariate EWMA control chart via multiple testing. *Journal of Process Control, 26*, 51–55. http://dx.doi.org/10.1016/j.jprocont.2015.01.007

Patel, A. K., & Divecha, J. (2011). Modified exponentially weighted moving average (EWMA) control chart for an analytical process data. *Journal of Chemical Engineering and Materials Science, 2*(1), 12–20.

Pignatiello, J. J., & Perry, M. B. (2011). Estimating the time of step change with Poisson CUSUM and EWMA control charts. *International Journal of Production Research, 49*(10/11), 2857–2871. doi:10.1080/00207541003690082

Roberts, S. W. (1959). Control Chart Tests Based on Geometric Moving Averages. *Technometrics, 1*(3), 239–250.

Radhakrishnan, R., & Sivakumaran, P. (2008). Construction of six sigma repetitive group sampling plans. *International Journal of Mathematics & Computation, 1*(N08), 75–83.

Shamsuzzaman, M., & Wu, Z. (2012). Design of EWMA control chart for minimizing the proportion of defective units. *International Journal of Quality & Reliability Management, 29*(8), 953–969. doi:10.1108/02656711211270379

Shankar, G. (1992). Procedures and tables for construction and selection of repetitive group sampling plans for three-attribute classes. *International Journal of Quality & Reliability Management, 9*(1), 27–37.

Shankar, G., & Mohapatra, B. (1993). GERT analysis of conditional repetitive group sampling plan. *International Journal of Quality & Reliability Management, 10*(2), 50–62.

Sherman, R. E. (1965). Design and evaluation of a repetitive group sampling plan. *Technometrics, 7*(1), 11–21.

Soudararajan, V., & Ramasamy, M. (1984). Designing repetitive group sampling (RGS) plan indexed by AQL and LQL. *ZAPQR Transactions, 9*(1), 9–14.

Sua, Y., Shub, L., & Tsui, K. L. (2011). Adaptive EWMA procedures for monitoring processes subject to linear drifts. *Computational Statistics and Data Analysis, 55,* 2819–2829.

Su-Fen, Y., Wen-Chi, T., Tzee-Ming, H., Chi-Chin, Y., & Smiley, C. (2011). Monitoring process mean with a new EWMA control chart. *Production, 21*(2), 217–222.

Teh, S. Y., Khoo, M. B., & Wu, Z. (2011). A sum of squares double exponentially weighted moving average chart. *Computers & Industrial Engineering, 61*(4), 1173–1188.

Tien, F.-C., & Liu, C.-S. (2011). A single-featured EWMA-X control chart for detecting shifts in process mean and standard deviation. *Journal of Applied Statistics, 38*(11), 2575–2596.

Trietsch, D. (1998). *Statistical quality control: A loss minimization approach* (Vol. 10). Singapore: World Scientific Publishing Co. Pte Ltd.

Tukey, J. (1960). A survey of sampling from contaminated distributions. In *Contributions to probability and statistics: volume dedicated to harold hetelling.* Stanford, CA: Stanford University Press.

Woodall, W. H. (1983). The distribution of the run length of one-sided CUSUM procedures for continuous random variables. *Technometrics, 25*(3), 295–301.

Xie, M., Goh, T. N., & Kuralmani, V. (2012). *Statistical models and control charts for high-quality processes.* Norwell, MA: Springer Science & Business Media.

Yang, S.-F. (2013). Using a new VSI EWMA average loss control chart to monitor changes in the difference between the process mean and target and/or the process variability. *Applied Mathematical Modelling, 37*, 7973–7982.

Zhang, J., Li, Z., Chen, B., & Wang, Z. (2014). A new exponentially weighted moving average control chart for monitoring the coefficient of variation. *Computers & Industrial Engineering, 78*, 205–212.

Zou, C., & Tsung, F. (2010). Likelihood ratio-based distribution-free EWMA control charts. *Journal of Quality Technology, 42*(2), 174.

6

Sampling Schemes for Developing Control Charts

6.1 Single Sampling Scheme

The control chart technique is the most common and acceptable process monitoring and surveillance scheme because of its straightforwardness. However, this straightforward scheme for monitoring the process, if any assignable cause has been occurred, is shown in the posting of the quality characteristics on the two lines, i.e. upper control limit (UCL) and the lower control limit (LCL), which is very slow process of monitoring the process (Chandara, 2001; Cochran, 2007; Singh, Tailor, Singh, & Kim, 2007). Many sampling schemes have been introduced for refining the quick monitoring of the process, for example, double, multiple, sequential, and repetitive sampling schemes (Allen, 2006; A. F. Costa & Rahim, 2004; Govindaraju & Subramani, 1993; Oakland, 2008; Schilling, 1982; Sen, 1971).

Simple random sampling scheme is the most commonly used and the simplest sampling scheme. The assumptions to apply such type of sampling largely depend on the formation of the population units. When all the units in the population have the known but not necessarily to be equal probability of being selected in the sample, then such sampling is termed as simple random sampling. The instance of single sampling scheme can be used in a data of the form single sampling scheme by attributes and by variables (Aslam, Nazir, & Jun, 2015). A single random of size, say n, can be selected from the homogeneous population as $x_1, x_2, x_3, ..., x_n$. These samples are evaluated by the specific statistical techniques for deciding whether it is falling in in-control or out-of-control region (Singh & Vishwakarma, 2009).

Preamble: The wise selection of sampling from the manufacturing process leads to quick decision about the process under study. In this chapter the different sampling schemes have been discussed for the development of an efficient control chart scheme.

Introduction to Statistical Process Control, First Edition. Muhammad Aslam, Aamir Saghir, and Liaquat Ahmad.

Other sampling schemes such as double, multiple, and sequential sampling schemes allow for several consecutive samples to reach the decision for its in-control or out-of-control region (Aslam, Azam, Khan, & Jun, 2015). It is based on the specification range model described above. The single sampling is based on estimating μ from the sample mean $\bar{x}$, and the σ^2, if it is unknown, is estimated from the sample variance S^2 Rueda, Martinez, Arcos, and Munoz (2009).

In simple random sampling, also known as the traditional sampling scheme, we simply select a random sample of any size, may be a variable sample size, at fixed time intervals or variable time intervals.

Due to less efficiency in quick detecting the out-of-control process, a number of alternative sample selection techniques have been developed for the efficient monitoring of the control charts. This sampling scheme is easy to apply and easy to understand, and much has been discussed thoroughly in almost all standard quality control books (Xie, Goh, & Kuralmani, 2012).

6.2 Double Sampling Scheme

Statistical efficiency of the control chart was improved to monitor the process by the introduction of the DS $\bar{X}$ chart. Under the DS procedure, the whole of the sample of size $n_1 + n_2$ has been collected at a time, and n_1 sample is used to decide the process state (in-control or the out-of-control process). If we remained undecided about the process state, then we have to consume the n_2 sample for efficient monitoring of the process. The DS scheme increases the monitoring efficiency of the control chart scheme by reducing the time required in the single sampling scheme without increasing the false alarm rate and the in-control process monitoring rate. Motivated by the efficiency of the DS $\bar{X}$ chart, many researchers of the quality control have introduced the DS dispersion charts (DS S-chart), combined $\bar{X}$ and the S-chart, and the multiple multivariate charts that were presented for the efficient monitoring of the production process (A. Costa & Machado, 2008; Mason & Young, 2002).

The DS scheme has been introduced to detect the small shift in the production process by including a second sample while making the decision without any interruption. Two concepts are commonly employed for the DS schemes. The first one is the concept of variable sampling interval. This concept is used to increase the efficiency of the control chart for its quick monitoring. We select a sample from the production process to observe the process's nature of deterioration, and we select the next sample after a reasonable long time if it shows its measurement near the central line, and we select the next sample very instantly if the sample shows any indication of the process deterioration. The second concept toward efficient monitoring of the production process is the variable sample size. In variable sample size the same technique of variable sampling interval is employed for increasing the efficiency of the proposed technique.

The DS scheme uses both these ideas of the variable sampling intervals and the variable sample sizes (He, Grigoryan, & Sigh, 2002). In DS scheme the results from the first sample determine the interval and the size of the sample. Whenever a process approaches to the warning limits, the second sample becomes necessary to declare the process out of control (Machado & Costa, 2008).

There are two types of control charts applied to monitor the process efficiently. The first type of DS control chart was proposed by Croasdale (1974) in which the fate of the process is determined on the information obtained from two separate samples and the conformation is made only on second sample. Daudin (1992) introduced another type of DS control chart scheme, in which the process fate is decided by the results obtained by combining the information of both the samples at the draw of the second sample. Thus the obtained larger sample has smaller standard deviation.

The procedure of the DS plan introduced by Croasdale (1974) can be described as follows:

1) Select first sample of size n_1 (x_{1i}, $i = 1, 2, 3, ..., n_1$) and a second sample of size n_2 (x_{2i}, $i = 1, 2, 3, ..., n_2$) from a population with known mean μ_0 and standard deviation σ.

2) Find the mean of first sample $\overline{X}_1 = \sum_{i=1}^{n_1} x_{1i}/n_1$.

3) The first stage limits are $[-M_1, M_1]$. The process will be considered as in control if $\dfrac{\overline{X}_1 - \mu_0}{\sigma/\sqrt{n_1}}$ lies between $[-M_1, M_1]$; otherwise, we observe the second stage

by calculating the mean of the second sample $\overline{X}_2 = \sum_{i=1}^{n_1} x_{2i}/n_2$.

4) Set the limits of the second stage as $[-M_2, M_2]$. The process will be considered as out of control if $\dfrac{\overline{X}_2 - \mu_0}{\sigma/\sqrt{n_2}} = Z_1 < -M_2$ or if $\dfrac{\overline{X}_2 - \mu_0}{\sigma/\sqrt{n_2}} = Z_2 > M_2$; otherwise, the process is in control (Figure 6.1).

The type of control chart introduced by Daudin (1992) is different from that introduced by Croasdale (1974) on the basis of the warning limits that can be described as follows:

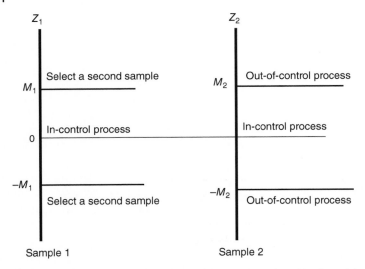

Figure 6.1 The scheme of double sampling plan introduced by Croasdale (1974) (Irianto & Juliani, 2010).

1) Select first sample of size n_1 (x_{1i}, $i = 1, 2, 3, ..., n_1$) and a second sample of size n_2 (x_{2i}, $i = 1, 2, 3, ..., n_2$) from a population with known mean μ_0 and standard deviation σ.

2) Find the mean of first sample $\overline{X}_1 = \sum_{i=1}^{n_1} x_{1i}/n_1$.

3) If Z_1 lies between $[-L_1, L_1]$, the process is declared as in control.

4) If Z_1 lies $< -L_1$ or $Z_1 > L_1$, the process is declared as out of control.

5) If Z_1 lies between $-L$ and $-L_1$ or L and L_1, then proceed to second sample with mean $\overline{X}_2 = \sum_{i=1}^{n_1} x_{2i}/n_2$.

6) Calculate the combine mean of two samples as $\overline{X} = (n_1\overline{X}_1 + n_2\overline{X}_2)/(n_1 + n_2)$ with standard deviation $\sigma/\sqrt{n_1 + n_2}$.

7) If $-L < Z_1 < -L_1$ or $L_1 < Z_1 < L$ and if $Z_{\text{combine}} = \dfrac{\overline{X} - \mu_0}{\sigma/\sqrt{n_1 + n_2}} < -L_2$ or $Z_{\text{combine}} = \dfrac{\overline{X} - \mu_0}{\sigma/\sqrt{n_1 + n_2}} > L_2$, then the process is declared as out of control; otherwise, the process is declared as in control (Figure 6.2).

The estimation of the process parameters is made by many available techniques in the literature, for example, optimum solution method, Monte Carlo simulation method, and integral equation method.

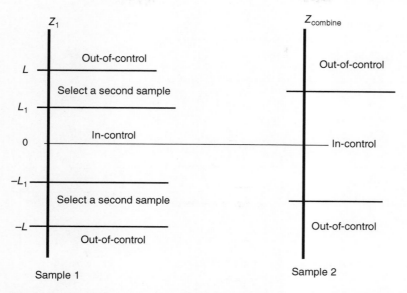

Figure 6.2 The scheme of DS plan introduced by Daudin (1992) (Irianto & Juliani, 2010).

6.3 Repetitive Sampling Scheme

The repetitive group sampling (RGS) is one of the important sampling scheme most commonly used in the literature of the quality control nowadays, which was introduced by Sherman (1965) for measuring the attribute data. This sampling scheme is more efficient with respect to the sample size as well as the desired protection than the single and DS schemes but less efficient than the sequential probability ratio sampling scheme. The RGS scheme is particularly more useful when used for the destructive sampling scenario or when the testing of samples is too costly (Aldosari, Aslam, & Jun, 2017). The operational procedure of the repetitive sampling is entirely different from the DS but similar to the sequential probability ration sampling. The RGS is quite simple to apply and to understand as compared with the sequential probability ratio sampling scheme (Figure 6.3).

For example, the RGS scheme can be explained by using the data given on page 232 of Montgomery (2009), and the developments of the UCL and the LCL can be explained in the following (Kenett, Zacks, & Amberti, 2013). The specification limits are 1.50 ± 1.50, and the estimated values from the observations are presented in Table 6.1.

Then two pairs of control limits can be constructed using $\overline{\overline{X}} = 1.5056$, $\bar{s} = 0.1315$, $T = 1.000$, and $C_p = 1.192$ (Adeoti & Olaomi, 2018):

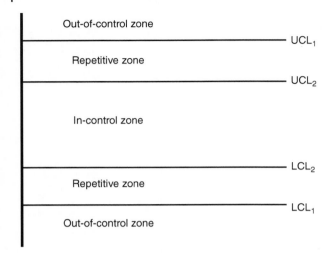

Figure 6.3 Graphical representation of the RGS scheme (Adeoti & Olaomi, 2018).

$$\text{UCL}_1 = \overline{\overline{X}} + k_1 \frac{T}{C_p},$$

$$= 1.5056 + 0.1492\left(\frac{1.000}{1.192}\right)$$

$$= 1.6308$$

$$\text{UCL}_2 = \overline{\overline{X}} + k_2 \frac{T}{C_p},$$

$$= 1.5056 + 0.0405\left(\frac{1.000}{1.192}\right)$$

$$= 1.5396$$

$$\text{LCL}_2 = \overline{\overline{X}} - k_2 \frac{T}{C_p},$$

$$= 1.5056 - 0.0405\left(\frac{1.000}{1.192}\right)$$

$$= 1.4716$$

$$\text{LCL}_1 = \overline{\overline{X}} - k_1 \frac{T}{C_p},$$

$$= 1.5056 - 0.1492\left(\frac{1.000}{1.192}\right)$$

$$= 1.3804$$

Table 6.1 Data for the hard bake process of flow width measurements (Montgomery, 2009, p. 232).

Sample number	Wafers						
	1	2	3	4	5	$\bar{x}$	R
1	1.3235	1.4128	1.6744	1.4573	1.6914	1.5119	0.3679
2	1.4314	1.3592	1.6075	1.4666	1.6109	1.4951	0.2517
3	1.4284	1.4871	1.4932	1.4324	1.5674	1.4817	0.1390
4	1.5028	1.6352	1.3841	1.2831	1.5507	1.4712	0.3521
5	1.5604	1.2735	1.5265	1.4363	1.6441	1.4882	0.3706
6	1.5955	1.5451	1.3574	1.3281	1.4198	1.4492	0.2674
7	1.6274	1.5064	1.8366	1.4177	1.5144	1.5805	0.4189
8	1.419	1.4303	1.6637	1.6067	1.5519	1.5343	0.2447
9	1.3884	1.7277	1.5355	1.5176	1.3688	1.5076	0.3589
10	1.4039	1.6697	1.5089	1.4627	1.522	1.5134	0.2658
11	1.4158	1.7667	1.4278	1.5928	1.4181	1.5242	0.3509
12	1.5821	1.3355	1.5777	1.3908	1.7559	1.5284	0.4204
13	1.2856	1.4106	1.4447	1.6398	1.1928	1.3947	0.4470
14	1.4951	1.4036	1.5893	1.6458	1.4969	1.5261	0.2422
15	1.3589	1.2863	1.5996	1.2497	1.5471	1.4083	0.3499
16	1.5747	1.5301	1.5171	1.1839	1.8662	1.5344	0.6823
17	1.368	1.7269	1.3957	1.5014	1.4449	1.4874	0.3589
18	1.4163	1.3864	1.3057	1.621	1.5573	1.4573	0.3153
19	1.5796	1.4185	1.6541	1.5116	1.7247	1.5777	0.3062
20	1.7106	1.4412	1.2361	1.382	1.7601	1.5060	0.5240
21	1.4371	1.5051	1.3485	1.567	1.488	1.4691	0.2185
22	1.4738	1.5936	1.6583	1.4973	1.472	1.5390	0.1863
23	1.5917	1.4333	1.5551	1.5295	1.6866	1.5592	0.2533
24	1.6399	1.5243	1.5705	1.5563	1.533	1.5648	0.1156
25	1.5797	1.3663	1.624	1.3732	1.6887	1.5264	0.3224

The operation of the RGS scheme based on the sampling mean can be explained in the following steps:

Step I: Let a random sample of size n be selected for the production process and calculate the sample mean as $\overline{X} = \sum\limits_{i=1}^{n} x_i/n$.

Step II: If $\overline{X} \geq UCL_1$ or $\overline{X} \leq LCL_1$ where UCL_1 and LCL_1 are called the outer control limits of the process, then the process is declared as out of control, and if $LCL_2 \leq \overline{X} \leq UCL_2$ where LCL_2 and UCL_2 are called the inner control limits of the process, then the process is declared as in control. Otherwise repeat the process starting from Step I.

The control chart performance can be evaluated by using the technique of average run length (ARL), which may be defined as the average number of samples before the process shows the out-of-control indication. For the estimation of the ARL of the process, the development of the control chart methodology using the RGS scheme can be developed as:

The four control limits of the RGS scheme can be written as follows:

$$UCL_1 = \overline{\overline{X}} + A_{21}\overline{R} \tag{6.1}$$

$$UCL_2 = \overline{\overline{X}} + A_{22}\overline{R} \tag{6.2}$$

$$LCL_2 = \overline{\overline{X}} - A_{22}\overline{R} \tag{6.3}$$

$$LCL_1 = \overline{\overline{X}} - A_{21}\overline{R} \tag{6.4}$$

where A_{21} and A_{22} are the control chart coefficients.

Let P_{in} denote the probability of in-control process for a single sample when the process mean μ is known, then

$$P_{in} = P\left(LCL_2 \leq \overline{X} \leq UCL_2\right) \tag{6.5}$$

$$P_{in} = \Phi\left(A_{22}d_2\sqrt{n}\right) - \Phi\left(-A_{22}d_2\sqrt{n}\right) \text{ and} \tag{6.6}$$

Let P_{out} denote the probability of out-of-control process for a single sample when the process mean μ is known, then

$$P_{out} = P\left(\overline{X} \geq UCL_1\right) + P\left(\overline{X} \leq LCL_2\right) \tag{6.7}$$

$$P_{out} = 1 - \Phi\left(A_{21}d_2\sqrt{n}\right) + \Phi\left(-A_{21}d_2\sqrt{n}\right) \tag{6.8}$$

Let P_{Rep} denote the probability of repetitive process for a single sample when the process mean μ is known, then

$$P_{Rep} = P\left(UCL_2 \leq \overline{X} \leq UCL_1\right) + P\left(LCL_1 \leq \overline{X} \leq LCL_2\right) \tag{6.9}$$

$$P_{Rep} = \Phi\left(A_{21}d_2\sqrt{n}\right) - \Phi\left(A_{22}d_2\sqrt{n}\right) + \Phi\left(-A_{22}d_2\sqrt{n}\right) - \Phi\left(-A_{21}d_2\sqrt{n}\right) \tag{6.10}$$

6.3.1 When a Shift of $\mu_1 = \mu + k\sigma$ Occurs in the Process

Let P_{in}^1 denote the probability of in-control process for a single sample when the process mean μ is known, then

$$P_{in}^1 = P(LCL_2 \leq \overline{X} \leq UCL_2) \mid \mu_1 = \mu + k\sigma \tag{6.11}$$

$$P_{in}^1 = \Phi\left(k\sqrt{n} + A_{22}d_2\sqrt{n}\right) - \Phi\left(k\sqrt{n} - A_{22}d_2\sqrt{n}\right) \tag{6.12}$$

Let P_{out}^1 denote the probability of out-of-control process for a single sample when the process mean μ is known, then

$$P_{out}^1 = P(\overline{X} \geq UCL_1) + P(\overline{X} \leq LCL_2) \mid \mu_1 = \mu + k\sigma \tag{6.13}$$

$$P_{out}^1 = 1 - \Phi\left(k\sqrt{n} + A_{21}d_2\sqrt{n}\right) + \Phi\left(k\sqrt{n} - A_{22}d_2\sqrt{n}\right) \tag{6.14}$$

Let P_{Rep}^1 denote the probability of in-control process for a repetitive sample when the process mean μ is known, then

$$P_{Rep}^1 = P(UCL_2 \leq \overline{X} \leq UCL_1) \mid \mu_1 = \mu + k\sigma + P(LCL_1 \leq \overline{X} \leq LCL_2) \\ \mid \mu_1 = \mu + k\sigma \tag{6.15}$$

$$P_{Rep}^1 = \Phi\left(k\sqrt{n} + A_{21}d_2\sqrt{n}\right) - \Phi\left(k\sqrt{n} - A_{22}d_2\sqrt{n}\right) + \Phi\left(k\sqrt{n} - A_{22}d_2\sqrt{n}\right) \\ - \Phi\left(k\sqrt{n} - A_{21}d_2\sqrt{n}\right) \tag{6.16}$$

Now the ARL formula for the shifted process can be described as (Table 6.2)

$$ARL_1 = \frac{1}{1 - P_{in}^1} \tag{6.17}$$

The values of the control chart coefficients $K_1 = 1.032$ and $K_2 = 0.587$ for $n = 3$ and the factor $d_2 = 1.693$ have been estimated by fixing the value of $ARL_0 = 370$ using the abovementioned Eqs. (6.1) through (6.4). Thus the control limits are calculated as

$$UCL_1 = 2476.30 + 1.032(397.52)$$
$$= 2886.5406$$
$$UCL_2 = 2476.30 + 0.0587(397.52)$$
$$= 27,090.64$$
$$LCL_2 = 2476.30 - 0.0587(397.52)$$
$$= 2242.95$$
$$LCL_1 = 2476.30 + 1.032(397.52)$$
$$= 2066.059$$

Table 6.2 Data for the calorific values (Elevli, 2006).

Sample number	Calorific values				
	1	2	3	$\bar{x}$	R
1	2698	3581	3020	3099.67	883
2	2471	2350	2520	2447.00	170
3	2283	2387	2679	2449.67	396
4	2598	3482	3051	3043.67	884
5	2943	2456	2361	2586.67	582
6	2236	2118	2500	2284.67	382
7	2427	2162	2184	2257.67	265
8	2697	3003	2428	2709.33	575
9	2206	2591	2565	2454.00	385
10	2540	2249	2430	2406.33	291
11	2714	2095	2280	2363.00	619
12	2352	2449	2584	2461.67	232
13	2296	2793	2827	2638.67	531
14	2279	2362	2204	2281.67	158
15	2411	2367	2393	2390.33	44
16	2151	2366	2097	2204.67	269
17	2130	2189	2557	2292.00	427
18	2330	2090	2054	2158.00	276
19	2040	2301	2591	2310.67	551
20	2444	2489	2677	2536.67	233
21	2719	2424	2736	2626.33	312
22	2506	2719	2343	2522.67	376
23	2506	2543	2241	2430.00	302

Figure 6.4 has been constructed using the mentioned control limits; thus a point above the UCL_2 or less than LCL_2 and/or above the UCL_1 or below the LCL_1 is declared as the out-of-control process. The process is declared as out of control because there are four observations falling out of control: two above the outer control limits and two below the inner control limits. Figure 6.5 also shows the out-of-control process as there are eight observations indicating this situation.

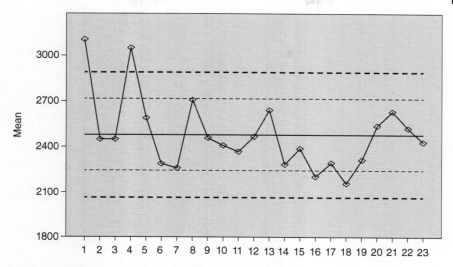

Figure 6.4 RGS control chart for calorific values (Ahmad, Aslam, & Jun, 2014).

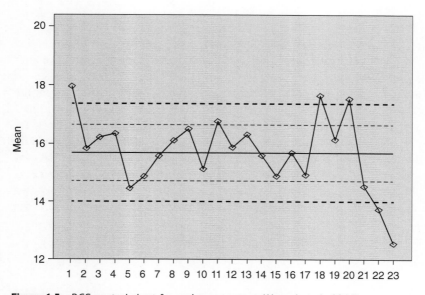

Figure 6.5 RGS control chart for moisture content (Ahmad et al., 2014).

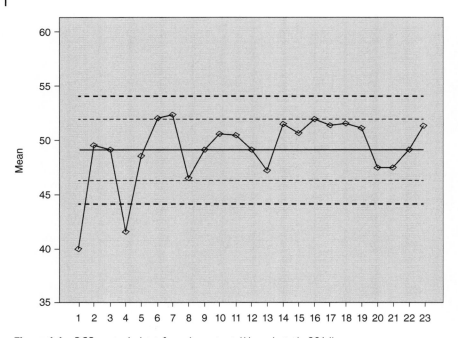

Figure 6.6 RGS control chart for ash content (Ahmad et al., 2014).

Figure 6.6 also shows out-of-control process as there are two observations falling above the inner control limit and two points falling below the outer control limit (Tables 6.3 and 6.4).

Normal probability distribution is the most commonly used distribution for the control chart literature. The abovementioned repetitive sampling has been utilized using the normal distribution. The scheme of RGS can also be developed for other probability distributions, for example, the gamma distribution (Balamurali, Jeyadurga, & Usha, 2016). This distribution is well known to fit the data for waiting time of life testing scenarios. Suppose the interested quality characteristic follows the gamma distribution. We have to develop two pairs of control limits with the object of the quickest monitoring of the out-of-control condition of the production process.

The cumulative distribution function of the gamma distribution is

$$P(T \le t) = 1 - \sum_{j=1}^{a-1} \frac{e^{-\frac{t}{b}}(t/b)^j}{j!} \qquad (6.18)$$

Table 6.3 Data for the moisture content values (Elevli, 2006).

| Sample number | Moisture contents | | | | |
	1	2	3	$\bar{x}$	R
1	17.2	18.41	18.24	17.95	1.21
2	16.27	15.58	15.58	15.81	0.69
3	16.11	15.96	16.58	16.22	0.62
4	14.68	18.68	15.7	16.35	4.00
5	15.06	14.55	13.7	14.44	1.36
6	14.68	14.87	15.07	14.87	0.39
7	14.6	15.47	16.67	15.58	2.07
8	15.43	16.56	16.35	16.11	1.13
9	15.84	16.17	17.51	16.51	1.67
10	14.86	14.93	15.54	15.11	0.68
11	19.42	15.6	15.34	16.79	4.08
12	15.76	15.27	16.5	15.84	1.23
13	17.62	16.09	15.2	16.30	2.42
14	15.87	15.65	15.23	15.58	0.64
15	15.02	14.71	14.91	14.88	0.31
16	15.23	15.65	16.21	15.70	0.98
17	15.6	14.78	14.37	14.92	1.23
18	17.37	18.83	16.75	17.65	2.08
19	17.54	15.53	15.25	16.11	2.29
20	20.01	18.18	14.43	17.54	5.58
21	14.51	14.92	14.11	14.51	0.81
22	14.18	13.61	13.42	13.74	0.76
23	13.24	12.34	12.07	12.55	1.17

where T is a random variable from the gamma distribution with two parameters interested quality characteristic a (shape) and b (scale). Wilson and Hilferty (1931) suggested that the random variable T can be transformed to $T^* = T^{1/3}$, which is approximately normally distributed with and variance given as

$$\mu_{T^*} = \frac{b^{1/3}\,\Gamma(a + 1/3)}{\Gamma(a)} \tag{6.19}$$

Table 6.4 Data for the ash content values (Elevli, 2006).

Sample number	Ash contents				
	1	2	3	$\bar{x}$	R
1	44.6	34.64	40.74	39.99	9.96
2	48.9	51.09	48.75	49.58	2.34
3	51.05	50.25	46.15	49.15	4.90
4	48.59	35.04	41.1	41.58	13.55
5	44.53	50.07	51.04	48.55	6.51
6	52.89	53.76	49.51	52.05	4.25
7	50.46	53.9	52.76	52.37	3.44
8	47.08	42.46	50.05	46.53	7.59
9	52.61	47.59	47.18	49.13	5.43
10	49.27	52.4	50.05	50.57	3.13
11	44.29	54.03	53.01	50.44	9.74
12	50.33	49.67	47.32	49.11	3.01
13	50.54	45.06	46.09	47.23	5.48
14	51.52	50.81	52.06	51.46	1.25
15	50.21	51.04	50.75	50.67	0.83
16	52.78	50.09	52.96	51.94	2.87
17	53.25	52.43	48.51	51.40	4.74
18	49.94	51.42	53.42	51.59	3.48
19	53.58	51.42	48.32	51.11	5.26
20	46.8	48.07	47.72	47.53	1.27
21	46.75	49.34	46.47	47.52	2.87
22	49.13	46.81	51.38	49.11	4.57
23	50.51	50.11	53.48	51.37	3.37

$$\sigma_{T^*} = \frac{b^{2/3}\,\Gamma(a + 2/3)}{\Gamma(a)} - \mu_{T^*}^2, \tag{6.20}$$

respectively.

Using the RGS scheme the control chart for the transformed gamma distribution can be developed as

Step I: Let we select a sample randomly and calculate its quality characteristic T. Then we calculate $T^* = T^{1/3}$.

Step II: If $T^* \geq UCL_1$ or $T^* \leq LCL_1$, then the process is declared as out of control. If $LCL_2 \leq T^* \leq UCL_2$, then the process is declared as in control. Otherwise, go to Step I and repeat the process.

Two pairs of control limits with two control coefficients K_1 and K_2 for the RGS scheme can be constructed as

$$LCL_1 = \mu_{T^*} - K_1\sigma_{T^*} \tag{6.21}$$

$$= \frac{b_0^{1/3}\Gamma(a + 1/3)}{\Gamma(a)} - K_1\sqrt{\frac{b_0^{2/3}\Gamma(a + 2/3)}{\Gamma(a)} - \mu_{T^*}^2} \tag{6.22}$$

$$UCL_1 = \mu_{T^*} + K_1\sigma_{T^*} \tag{6.23}$$

$$= \frac{b_0^{1/3}\Gamma(a + 1/3)}{\Gamma(a)} + K_1\sqrt{\frac{b_0^{2/3}\Gamma(a + 2/3)}{\Gamma(a)} - \mu_{T^*}^2} \tag{6.24}$$

Likewise, the inner control limits are

$$LCL_2 = \mu_{T^*} - K_1\sigma_{T^*} \tag{6.25}$$

$$= \frac{b_0^{1/3}\Gamma(a + 1/3)}{\Gamma(a)} - K_2\sqrt{\frac{b_0^{2/3}\Gamma(a + 2/3)}{\Gamma(a)} - \mu_{T^*}^2} \tag{6.26}$$

$$UCL_2 = \mu_{T^*} + K_1\sigma_{T^*} \tag{6.27}$$

$$= \frac{b_0^{1/3}\Gamma(a + 1/3)}{\Gamma(a)} + K_2\sqrt{\frac{b_0^{2/3}\Gamma(a + 2/3)}{\Gamma(a)} - \mu_{T^*}^2} \tag{6.28}$$

The coefficients K_1 and K_2 have been estimating using the following equations for specific values of ARL:

$$LL_1 = \frac{\Gamma(a + 1/3)}{\Gamma(a)} - K_1\sqrt{\frac{\Gamma(a + 2/3)}{\Gamma(a)} - \left(\frac{\Gamma(a + 1/3)}{\Gamma(a)}\right)^2} \tag{6.29}$$

$$UL_1 = \frac{\Gamma(a + 1/3)}{\Gamma(a)} + K_1\sqrt{\frac{\Gamma(a + 2/3)}{\Gamma(a)} - \left(\frac{\Gamma(a + 1/3)}{\Gamma(a)}\right)^2} \tag{6.30}$$

$$LL_2 = \frac{\Gamma(a + 1/3)}{\Gamma(a)} - K_2\sqrt{\frac{\Gamma(a + 2/3)}{\Gamma(a)} - \left(\frac{\Gamma(a + 1/3)}{\Gamma(a)}\right)^2} \tag{6.31}$$

$$UL_2 = \frac{\Gamma(a + 1/3)}{\Gamma(a)} + K_2\sqrt{\frac{\Gamma(a + 2/3)}{\Gamma(a)} - \left(\frac{\Gamma(a + 1/3)}{\Gamma(a)}\right)^2} \tag{6.32}$$

The following is the hypothetical data of 30 observations generated from the gamma distribution to construct the control limits of the above scheme (Table 6.5).

Table 6.5 The hypothetical 30 observations from the gamma distribution (Khan, Aslam, Ahmad, & Jun, 2017).

Sample number	T	Sample number	T
1	0.3557	16	0.1662
2	0.8709	17	1.0774
3	1.1496	18	4.0056
4	1.7524	19	0.6820
5	2.9577	20	0.8878
6	2.3009	21	0.0597
7	0.3100	22	0.8059
8	0.8291	23	0.7271
9	2.1613	24	0.6041
10	1.5632	25	1.6952
11	0.4296	26	0.6740
12	0.8914	27	0.8610
13	0.1014	28	0.3215
14	0.1210	29	0.5248
15	1.3688	30	0.0329

Table 6.6 has been developed for the transformed values with the transformation $T^* = T^{1/3}$.

The four control limits are constructed as $LCL_1 = -0.0223$, $LCL_2 = 0.0199$, $UCL_2 = 1.7660$, and $UCL_1 = 1.8082$. The posting of the data and the control limits has shown in Figure 6.7.

The abovementioned scheme of control chart for the gamma distribution has been applied on the data of the urinary tract infection given by Santiago and Smith (2013). The transformed data are provided in Table 6.7 and Figure 6.8.

6.4 Mixed Sampling Scheme

Generally, control charts have been constructed on the two major types of the quality characteristics: attribute control charts and the variable control charts. The first type of control charts is applied when the interested quality characteristic cannot be

Table 6.6 Transformation of the hypothetical 30 observations from the gamma distribution (Khan et al., 2017).

Sample number	T^*	Sample number	T^*
1	0.7086	16	0.5498
2	0.9549	17	1.0251
3	1.0475	18	1.5881
4	1.2056	19	0.8802
5	1.4354	20	0.9611
6	1.3201	21	0.3910
7	0.6768	22	0.9306
8	0.9394	23	0.8992
9	1.2929	24	0.8453
10	1.1605	25	1.1923
11	0.7545	26	0.8767
12	0.9624	27	0.9513
13	0.4663	28	0.6850
14	0.4946	29	0.8066
15	1.1103	30	0.3207

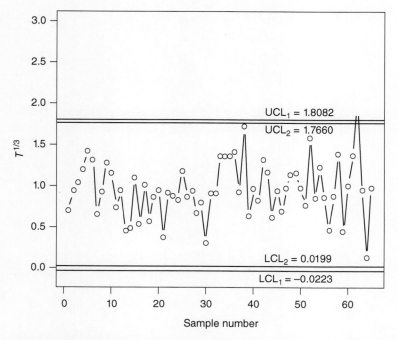

Figure 6.7 RGS control chart for the hypothetical data (Khan et al., 2017).

Table 6.7 Transformation of the urinary tract infection from Santiago and Smith (2013) (Khan et al., 2017).

Sample number	T^*	Sample number	T^*
1	1.5989	31	1.7503
2	1.9088	32	1.2716
3	1.5430	33	1.7535
4	1.8190	34	1.7035
5	1.8835	35	1.6461
6	1.4631	36	1.5905
7	1.2838	37	1.5605
8	2.1045	38	1.4286
9	1.7457	39	1.9444
10	1.7651	40	2.1286
11	1.8374	41	1.8089
12	1.3895	42	2.2751
13	1.8341	43	1.7196
14	1.5507	44	1.6306
15	1.5522	45	1.6914
16	1.8457	46	1.2228
17	1.5502	47	1.3023
18	1.4127	48	2.3409
19	1.2637	49	1.8352
20	1.5467	50	1.7123
21	2.3554	51	1.8181
22	2.0375	52	1.8539
23	1.4223	53	1.6534
24	1.5916	54	1.4182
25	1.5047	55	2.0628
26	2.1053	56	1.7667
27	1.8844	57	1.5866
28	1.5275	58	1.6654
29	1.7886	59	1.7574
30	1.4014	60	2.1864

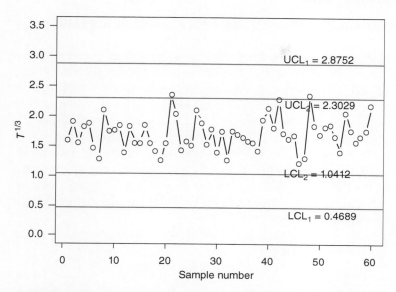

Figure 6.8 RGS control chart for the urinary tract infection from Santiago and Smith (2013).

measured numerically, which is labeled as good/bad, yes/no, passed/failed, etc. The other type of quality characteristic is the measurement such as height, weight, and length, which provides more information as compared with the attribute data. In general, more samples are required for the attribute type of data as compared to the variable type to reach any decision (Soundararajan & Vijayaraghavan, 1990). The monitoring ability of the developed control chart technique can be enhanced by utilizing mixture of the quality characteristic (Trietsch, 1998). This type of sampling scheme is known as the MS scheme.

6.4.1 Mixed Control Chart Using Exponentially Weighted Moving Average (EWMA) Statistics

In this section we designed two mixed control charts for the quality characteristic that follows normal probability distribution. This technique begins with monitoring of the number of nonconforming units and then switches to a different variable chart as EWMA statistic with the condition that the decision on the first technique is pending (Aslam, Khan, Aldosari, & Jun, 2016).

6.5 Mixed Control Chart Using Process Capability Index

The concept of process capability index is related with the study of capability of a process to produce the goods and services within the specification limits (Aslam, Azam, & Jun, 2013). This term of process capability index specifies the process to perform within the in-control limits. The comparison of different processes about how much a process varies from the natural limits to specification limits can be examined for its effectiveness. This comparison is made by forming the ratio of the variation in the specification limits to the variation of the process with six standard deviation units (Yan, Liu, & Dong, 2016). A process in which all the observations fall inside the specification limits is said to be a capable process. In simple words, it measures producer's capability to produce an item within the customer's tolerance range.

The formulation of the scheme based on the process capability index C_{pk} can be described in the following steps:

Step I: Let a subgroup of size n is selected from the process. Find d where d is the number of nonconforming items in the sample.

Step II: If $d > \text{UCL}_1$ or $d < \text{UCL}_1$, declare the process as out of control. If $\text{LCL}_2 \leq d \leq \text{UCL}_2$, then declare the process as in control; otherwise go to Step III.

Step III: Take a random sample of size n from the lot. Find the sample mean $\bar{X} = \sum_{i=1}^{n} x_i / n$ and the sample variance $S^2 = \left[\sum_{i=1}^{n} (x_i - \bar{X})^2 / {n-1} \right]$. Then compute $\hat{C}_{pk} = \min \left\{ \dfrac{\text{USL} - \bar{X}}{3S}, \dfrac{\bar{X} - \text{LSL}}{3S} \right\}$.

Step IV: If $\hat{C}_{pk} < \text{UCL}_3$, the process will be declared as out of control; otherwise the process is declared as in control.

The three pairs of control limits for the MS scheme can be developed as

$$\text{UCL}_1 = np_0 + K_1 \sqrt{np_0(1 - p_0)} \tag{6.33}$$

$$\text{LCL}_1 = \max \left[0, np_0 - K_1 \sqrt{np_0(1 - p_0)} \right] \tag{6.34}$$

$$\text{UCL}_2 = np_0 + K_2 \sqrt{np_0(1 - p_0)} \tag{6.35}$$

$$\text{LCL}_2 = \max \left[0, np_0 - K_2\sqrt{np_0(1-p_0)} \right] \tag{6.36}$$

where K_1 and K_2 are the control chart coefficients. Then the control limits for the variable C_{pk} can be constructed as

$$\text{UCL}_3 = \mu_{C_{\text{pk}}} + K_3\sigma_{C_{\text{pk}}} \tag{6.37}$$

where K_3 is the control limit coefficient of the third limit.

To calculate the ARL of the in-control and out-of-control process, the following equations can be developed:

$$P_{\text{in}} = P(\text{LCL}_2 \le d \le \text{UCL}_2) + \{P(\text{UCL}_2 \le d \le \text{UCL}_1)$$

$$+ P(\text{LCL}_1 \le d \le \text{LCL}_2)\} \times P(\text{LCL}_3 \le \hat{C}_{\text{pk}} \le \text{UCL}_3). \tag{6.38}$$

We define

$$f_1 = P(\text{LCL}_2 \le d \le \text{UCL}_2) \tag{6.39}$$

$$f_2 = P(\text{UCL}_2 \le d \le \text{UCL}_1) + P(\text{LCL}_1 \le d \le \text{LCL}_2) \tag{6.40}$$

$$f_3 = P(\hat{C}_{\text{pk}} < \text{UCL}_3) \tag{6.41}$$

$$= 1 - P(\hat{C}_{\text{pk}} \ge \text{UCL}_3) \tag{6.42}$$

Then we can write

$$P_{\text{in}} = f_1 + f_2 f_3 \tag{6.43}$$

$$f_{10} = P(\text{LCL}_2 \le d \le \text{UCL}_2 \mid p_0) = \sum_{d = \lfloor \text{LCL}_2 \rfloor + 1}^{\lfloor \text{UCL}_2 \rfloor} \binom{n}{d} p_0^d (1-p_0)^{n-d} \tag{6.44}$$

$$f_{20} = P(\text{UCL}_2 \le d \le \text{UCL}_1 \mid p_0) + P(\text{LCL}_1 \le d \le \text{LCL}_2 \mid p_0) \tag{6.45}$$

$$f_{30} = \left\{ \Phi\left((Z_{pu} - 3(\text{UCL}_3))\sqrt{\frac{n}{1 + 9(\text{UCL}_3)^2/2}} \right) \right.$$

$$\left. - \Phi\left(-(Z_{pl} - 3(\text{UCL}_3))\sqrt{\frac{n}{1 + 9(\text{UCL}_3)^2/2}} \right) \right\} \tag{6.46}$$

The ARL of the in-control process can be stated as

$$\text{ARL}_0 = \frac{1}{1 - P_{\text{in},0}} = \frac{1}{1 - (f_{10} + f_{20}f_{30})} \tag{6.47}$$

$$P_0 = P(X > \text{USL} \mid m) = 1 - \Phi\left(\frac{\text{USL} - m}{\sigma}\right) \tag{6.48}$$

When the process is out of control, the fraction nonconforming will be

$$P_1 = P(X > \text{USL} \mid m + c\sigma) = 1 - \Phi\left(\frac{\text{USL} - m - c\sigma}{\sigma}\right) = 1 - \Phi\left(\frac{\text{USL} - m}{\sigma} - c\right) \tag{6.49}$$

$$f_{11} = P(\text{LCL}_2 \le d \le \text{UCL}_2 \mid p_1) = \sum_{d = \lfloor \text{LCL}_2 \rfloor + 1}^{\lfloor \text{UCL}_2 \rfloor} \binom{n}{d} p_1^d (1 - p_1)^{n - d} \tag{6.50}$$

$$f_{21} = \sum_{d = \lfloor \text{LCL}_2 \rfloor + 1}^{\lfloor \text{UCL}_1 \rfloor} \binom{n}{d} p_1^d (1 - p_1)^{n - d} + \sum_{d = \lfloor \text{LCL}_1 \rfloor + 1}^{\lfloor \text{UCL}_2 \rfloor} \binom{n}{d} p_1^d (1 - p_1)^{n - d} \tag{6.51}$$

$$f_{31} = \left\{ \Phi\left((Z_{p_1 \text{U}} - 3(\text{UCL}_3)) \sqrt{\frac{n}{1 + 9(\text{UCL}_3)^2 / 2}} \right) \right.$$

$$\left. - \Phi\left(-(Z_{p_1 \text{L}} - 3(\text{UCL}_3)) \sqrt{\frac{n}{1 + 9(\text{UCL}_3)^2 / 2}} \right) \right\} \tag{6.52}$$

The ARL of the shifted process will be computed as

$$\text{ARL}_1 = \frac{1}{1 - (f_{11} + f_{21}f_{31})} \tag{6.53}$$

Using the abovementioned equations the estimation of the coefficients of the control chart has been given in Table 6.8 through 6.12 using the simulation approach for different process settings. For computing the detection ability of the abovementioned scheme, we used $p = 0.05, 0.10, 0.30,$ and 0.50, with high values of the in-control process $\text{ARL}_0 = 100, 200, 300,$ and 370, with different shift levels $0.01, 0.02, 0.05, 0.10, 0.15, 0.20, 0.25, 0.30, 0.40, 0.50, 0.60,$ and 0.70.

Table 6.8 Values of ARL for $p = 0.05$.

		$p = 0.05$		
n	195	168	128	141
k_1	2.6569	2.8961	3.4076	3.1161
k_2	1.1900	1.5952	0.1336	0.9866
UCL_3	0.4147	0.8318	0.4529	0.8779
m		ARLs		
0	100.15	201.27	301.07	370.13
0.01	83.80	167.41	281.40	330.24
0.02	70.06	139.11	257.14	289.08
0.05	41.31	80.41	177.22	180.64
0.1	18.32	34.23	82.03	77.23
0.15	9.03	16.03	37.91	34.55
0.2	4.98	8.30	18.68	16.79
0.25	3.06	4.76	9.95	8.93
0.3	2.09	3.01	5.74	5.20
0.4	1.29	1.60	2.42	2.30
0.5	1.06	1.16	1.40	1.41
0.6	1.01	1.03	1.07	1.11
0.7	1.00	1.00	1.00	1.02
0.8	1.00	1.00	1.00	1.00
0.9	1.00	1.00	1.00	1.00
1.0	1.00	1.00	1.00	1.00

Table 6.9 Values of ARL for $p = 0.10$.

		$p = 0.10$		
n	53	72	90	64
k_1	3.0629	3.5863	3.0032	3.3828
k_2	1.7464	0.2082	0.5768	0.5989
UCL_3	0.4036	0.4020	0.3589	0.3588
m		ARLs		
0	100.46	200.04	301.35	371.15
0.01	95.07	172.67	251.78	350.46
0.02	88.91	148.36	210.59	324.56
0.05	68.62	92.97	124.72	234.87
0.1	39.75	43.11	55.05	116.50

(*Continued*)

Table 6.9 (Continued)

n				
			p = 0.10	
	53	72	90	64
0.15	22.14	21.05	26.25	55.79
0.2	12.58	10.98	13.55	27.73
0.25	7.48	6.16	7.57	14.60
0.3	4.71	3.73	4.57	8.21
0.4	2.25	1.73	2.10	3.21
0.5	1.39	1.10	1.31	1.68
0.6	1.20	1.10	1.06	1.16
0.7	1.11	1.11	1.00	1.00
0.8	1.05	1.05	1.00	1.02
0.9	1.02	1.01	1.00	1.01
1.0	1.00	1.00	1.00	1.00

Table 6.10 Values of ARL for $p = 0.20$.

n				
			p = 0.20	
	69	34	58	87
k_1	2.6094	2.7792	3.0545	3.1995
k_2	0.8550	1.2454	0.9416	1.0380
UCL_3	0.6528	0.7403	1.0444	0.5878
m			ARLs	
0	100.11	200.57	300.13	370.61
0.01	93.47	183.88	258.63	360.53
0.02	85.62	167.52	221.96	338.45
0.05	60.84	123.48	139.18	236.95
0.1	31.18	71.76	65.44	101.79
0.15	16.31	42.00	32.73	43.02
0.2	9.15	25.34	17.56	19.74
0.25	5.54	15.90	10.11	10.02
0.3	3.63	10.39	6.24	5.63
0.4	1.93	5.03	2.91	2.39
0.5	1.33	2.86	1.74	1.43
0.6	1.10	1.88	1.27	1.12
0.7	1.02	1.42	1.09	1.02
0.8	1.00	1.19	1.02	1.00
0.9	1.00	1.08	1.00	1.00
1	1.00	1.03	1.00	1.00

Table 6.11 Values of ARL for $p = 0.30$.

			$p = 0.30$	
n	58	34	43	23
k_1	3.0601	2.8461	3.6074	2.9864
k_2	1.5026	0.2485	1.7483	1.0098
UCL_3	0.2511	0.1071	0.2318	0.1160
m			ARLs	
0	100.14	200.09	300.68	370.06
0.01	88.95	180.11	266.22	334.83
0.02	78.37	161.32	233.89	302.01
0.05	51.62	113.98	152.79	219.03
0.1	24.28	63.30	70.42	126.75
0.15	11.66	36.15	32.63	74.21
0.2	6.06	21.52	16.05	44.51
0.25	3.48	13.41	8.55	27.49
0.3	2.22	8.74	4.96	17.53
0.4	1.21	4.24	2.14	7.90
0.5	1.26	2.44	1.30	4.11
0.6	1.21	1.65	1.33	2.47
0.7	1.10	1.28	1.28	1.70
0.8	1.03	1.10	1.17	1.32
0.9	1.01	1.03	1.08	1.13
1.0	1.00	1.01	1.03	1.04

Table 6.12 Values of ARL for $p = 0.50$.

			$p = 0.50$	
n	18	19	31	20
k_1	2.5333	2.9179	2.8373	3.0761
k_2	0.0704	0.1421	0.2357	0.1409
UCL_3	0.6027	0.6940	0.7254	0.7618
m			ARLs	
0	100.03	200.46	300.24	370.35
0.01	100.99	200.81	297.60	369.23
0.02	101.22	199.27	289.84	363.94
0.05	97.56	184.50	244.28	327.16

(*Continued*)

Table 6.12 (Continued)

	p = 0.50			
n	18	19	31	20
0.1	80.81	140.74	152.00	234.58
0.15	60.29	97.43	88.00	153.51
0.2	42.99	65.51	51.55	98.63
0.25	30.40	44.24	31.28	64.13
0.3	21.70	30.40	19.76	42.63
0.4	11.67	15.37	8.90	20.34
0.5	6.83	8.54	4.68	10.75
0.6	4.35	5.20	2.83	6.26
0.7	3.00	3.45	1.94	3.99
0.8	2.22	2.47	1.49	2.77
0.9	1.75	1.90	1.25	2.07
1.0	1.46	1.55	1.12	1.65

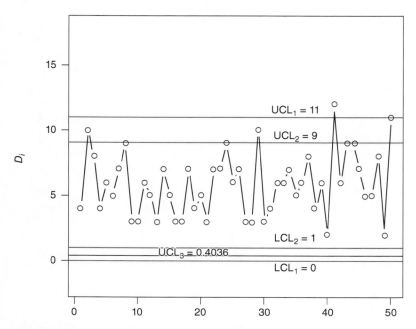

Figure 6.9 Control chart for simulated data (Aslam, Khan, Ahmad, Jun, & Hussain, 2017).

6.5.1 Analysis Through Simulation Approach

An R language statistical software was used to determine whether the developed process meets or does not meet the specific requirements using the abovementioned equations. A large number of samples were generated from the normal distribution with $p = 0.10$. Totally, 30 observations were generated in which the first 20 observations were generated using the in-control process condition and the next 30 observations were generated with the condition of a shift of $c = 0.15$. The simulated data have been plotted on the graph given in Figure 6.9.

References

Adeoti, O. A., & Olaomi, J. O. (2018). Capability index based control chart for monitoring process mean using repetitive sampling. *Communications in Statistics – Theory and Methods*, *47*(2), 493–507.

Ahmad, L., Aslam, M., & Jun, C.-H. (2014). Coal quality monitoring with improved control charts. *European Journal of Scientific Research*, *125*(2), 427–434.

Aldosari, M. S., Aslam, M., & Jun, C.-H. (2017). A new attribute control chart using multiple dependent state repetitive sampling. *IEEE Access*, *5*, 6192–6197.

Allen, T. T. (2006). *Introduction to engineering statistics and six sigma*. London: Springer-Verlag.

Aslam, M., Azam, M., & Jun, C.-H. (2013). Multiple dependent state sampling plan based on process capability index. *Journal of Testing and Evaluation*, *41*(2), 340–346.

Aslam, M., Azam, M., Khan, N., & Jun, C.-H. (2015). A control chart for an exponential distribution using multiple dependent state sampling. *Quality & Quantity*, *49*(2), 455–462. doi:10.1007/s11135-014-0002-2

Aslam, M., Khan, N., Ahmad, L., Jun, C.-H., & Hussain, J. (2017). A mixed control chart using process capability index. *Sequential Analysis*, *36*(2), 278–289.

Aslam, M., Khan, N., Aldosari, M. S., & Jun, C.-H. (2016). Mixed control charts using EWMA statistics. *IEEE Access*, *4*, 8286–8293.

Aslam, M., Nazir, A., & Jun, C.-H. (2015). A new attribute control chart using multiple dependent state sampling. *Transactions of the Institute of Measurement and Control*, *37*(4), 569–576.

Balamurali, S., Jeyadurga, P., & Usha, M. (2016). Designing of Bayesian multiple deferred state sampling plan based on Gamma–Poisson distribution. *American Journal of Mathematical and Management Sciences*, *35*(1), 77–90.

Chandara, M. J. (2001). *Statistical quality control*. Boca Raton, FL: CRC Press LLC.

Cochran, W. G. (2007). *Sampling techniques*. Wiley.

Costa, A., & Machado, M. (2008). Bivariate control charts with double sampling. *Journal of Applied Statistics*, *35*(7), 809–822.

Costa, A. F., & Rahim, M. (2004). Joint – X and R Charts with two-stage samplings. *Quality and Reliability Engineering International, 20*(7), 699–708.

Croasdale, R. (1974). Control charts for a double-sampling scheme based on average production run lengths. *International Journal of Production Research, 12*(5), 585–592.

Daudin, J. J. (1992). Double sampling charts. *Journal of Quality Technology, 24*(2), 78–87.

Elevli, S. (2006). Coal quality control with control charts. *Coal Preparation, 26*(4), 181–199.

Govindaraju, K., & Subramani, K. (1993). Selection of multiple deferred (dependent) state sampling plans for given acceptable quality level and limiting quality level. *Journal of Applied Statistics, 20*(3), 423–428.

He, D., Grigoryan, A., & Sigh, M. (2002). Design of double- and triple-sampling X-bar control charts using genetic algorithms. *International Journal of Production Research, 40*(6), 1387–1404.

Irianto, D., & Juliani, A. (2010). A two control limits double sampling control chart by optimizing producer and customer risks. *ITB Journal of Engineering Science, 42*(2), 165–178.

Kenett, R., Zacks, S., & Amberti, D. (2013). *Modern industrial statistics: With applications in R, MINITAB and JMP*. United Kingdom: Wiley.

Khan, N., Aslam, M., Ahmad, L., & Jun, C.-H. (2017). A control chart for gamma distributed variables using repetitive sampling scheme. *Pakistan Journal of Statistics and Operation Research, XIII*(1), 47–61.

Machado, M. A., & Costa, A. F. (2008). The double sampling and the EWMA charts based on the sample variances. *International Journal of Production Economics, 114*(1), 134–148.

Mason, R. L., & Young, J. C. (2002). *Multivariate statistical process control with industrial applications* (Vol. 9). Philadelphia, PA: Siam.

Montgomery, D. C. (2009). *Introduction to statistical quality control* (6th ed.). New York: Wiley.

Oakland, J. S. (2008). *Statistical process control* (6th ed.). Oxford: Butterworth-Heinemann.

Rueda, M., Martinez, S., Arcos, A., & Munoz, J. (2009). Mean estimation under successive sampling with calibration estimators. *Communications in Statistics – Theory and Methods, 38*(6), 808–827.

Santiago, E., & Smith, J. (2013). Control charts based on the exponential distribution: adapting runs rules for the t chart. *Quality Engineering, 25*(2), 85–96.

Schilling, E. G. (1982). *Acceptance sampling in quality control*. Baca Raton, FL: Chapman & Hall/CRC, Taylor & Francis Group.

Sen, A. (1971). Successive sampling with two auxiliary variables. *Sankhyā: The Indian Journal of Statistics, Series B*, 371–378.

Sherman, R. E. (1965). Design and evaluation of a repetitive group sampling plan. *Technometrics*, *7*(1), 11–21.

Singh, H. P., Tailor, R., Singh, S., & Kim, J.-M. (2007). Quantile estimation in successive sampling. *Journal of the Korean Statistical Society*, *36*(4), 543–556.

Singh, H. P., & Vishwakarma, G. K. (2009). A general procedure for estimating population mean in successive sampling. *Communications in Statistics – Theory and Methods*, *38*(2), 293–308.

Soundararajan, V., & Vijayaraghavan, R. (1990). Construction and selection of multiple dependent (deferred) state sampling plan. *Journal of Applied Statistics*, *17*(3), 397–409. doi:10.1080/02664769000000012

Trietsch, D. (1998). *Statistical quality control: A loss minimization approach* (Vol. 10). Singapore: World Scientific Publishing Co. Pte Ltd.

Wilson, E. B., & Hilferty, M. M. (1931). The distribution of chi-square. *Proceedings of the National Academy of Sciences*, *17*(12), 684–688.

Xie, M., Goh, T. N., & Kuralmani, V. (2012). *Statistical models and control charts for high-quality processes*. Norwell, MA: Springer Science & Business Media.

Yan, A., Liu, S., & Dong, X. (2016). Designing a multiple dependent state sampling plan based on the coefficient of variation. *SpringerPlus*, *5*(1), 1447. doi:10.1186/s40064-016-3087-3

7

Memory-Type Control Charts for Attributes

7.1 Exponentially Weighted Moving Average (EWMA) Control Charts for Attributes

It has been well discussed in the statistical process control (SPC) literature that EWMA control charts have better run length performance than Shewhart control charts for detecting small to moderate shifts in the parameter of interest. An EWMA chart dominates the Shewhart chart in detecting small to moderate shifts in terms of statistical performance. For a comprehensive overview of EWMA charts in attributes, see Woodall (2006). The general form for two-sided EWMA statistics, E_t, $t = 1, 2, ...$, is

$$E_t = \lambda V_t + (1 - \lambda)E_{t-1} \tag{7.1}$$

for random variable V_t, $t = 1, 2, ...$, and $E_0 = \mu_V$, the in-control mean of V. The parameter $\lambda (0 < \lambda < 1)$ is the weight assigned to be given to the most recent observation, and the EWMA chart is equivalent to the Shewhart chart for $\lambda = 1$. A weight in the range $0.05 \leq \lambda \leq 0.25$ is usually recommended (Montgomery, 2013, p. 423). The control limits for an EWMA chart at time t are

$$
\begin{aligned}
\text{LCL}(t) &= \mu_V - L_L \sigma_V \sqrt{\frac{\lambda\left[1 - (1 - \lambda)^{2t}\right]}{2 - \lambda}} \\
\text{CL} &= \mu_V \\
\text{UCL}(t) &= \mu_V + L_U \sigma_V \sqrt{\frac{\lambda\left[1 - (1 - \lambda)^{2t}\right]}{2 - \lambda}}
\end{aligned}
\tag{7.2}
$$

for $t = 1, 2, ...$, where L is a specified multiplier, σ_V is the standard deviation of random variable V, LCL is the lower control limit, CL is the central line, and UCL is the upper control limit. In the literature, asymptotic control limits for the EWMA chart are often used, where the term $[1 - (1 - \lambda)^{2t}]$ converges to 1 for $t \to \infty$.

Introduction to Statistical Process Control, First Edition. Muhammad Aslam, Aamir Saghir, and Liaquat Ahmad.

Generally, two-sided control limits defined in Eq. (7.2) are used. However, one-sided control limit can be used if the primary focus is to detect a change in a parameter in only one direction (either upward or downward).

7.1.1 Binomial EWMA Charts

Gan (1990) proposed a modified EWMA control chart for monitoring binomial counts. Suppose that a random sample of size n is taken from a production process with p as the in-control fraction of nonconforming product produced follows binomial distribution. The sample proportion of defects/fraction nonconforming can be defined as the ratio between the number of nonconforming and the total units in a sample size n, that is, $\hat{p} = \frac{x}{n}$. The fraction nonconforming EWMA statistic can be defined as

$$E_t = \lambda p_t + (1 - \lambda)E_{t-1}; \quad t = 1, 2, \ldots \tag{7.3}$$

where $E_0 = p_0$ is the initial EWMA assigned and p_t is the proportion of nonconforming products in the t-th sample.

The control limits for binomial EWMA control chart are

$$
\begin{aligned}
\text{LCL}(t) &= p_0 - L_L \sqrt{\frac{p_0(1-p_0)\,\lambda\left[1-(1-\lambda)^{2t}\right]}{n(2-\lambda)}} \\
\text{CL} &= p_0 \\
\text{UCL}(t) &= p_0 + L_U \sqrt{\frac{p_0(1-p_0)\,\lambda\left[1-(1-\lambda)^{2t}\right]}{n(2-\lambda)}}
\end{aligned}
\tag{7.4}
$$

where L_L and L_U are coefficients of EWMA control limits and usually taken equal, i.e. $L_L = L_U = L$. These control limits are known as **time-varying control limits**. When $t \to \infty$, the control limits of EWMA chart can be rewritten as the following:

$$
\begin{aligned}
\text{LCL} &= p_0 - L_L \sqrt{\frac{p_0(1-p_0)\lambda}{n(2-\lambda)}} \\
\text{CL} &= p_0 \\
\text{UCL} &= p_0 + L_U \sqrt{\frac{p_0(1-p_0)\,\lambda}{n(2-\lambda)}}
\end{aligned}
\tag{7.5}
$$

Random samples of products are usually taken at regular time intervals, and the number of nonconforming products is observed. An out-of-control alarm is issued when $E_t < h$, or $E_t < h$, otherwise the process is declared in control. The reference value h is required to determine based on the parameters of binomial distribution and at the given level of false alarm rate.

The average run length (ARL) and the probability function of the run length of the binomial EWMA control chart can be calculated using the Markov chain theory or Monte Carlo simulation. The modified EWMA control chart was demonstrated to be generally superior to the Shewhart control chart based on ARL consideration. Gan (1990) determined the ARL values for binomial EWMA chart using Markov chain approach. Table 7.1 gives ARL comparisons among binomial EWMA and upper-sided Shewhart charts for sample size 200 and $p = 0.10$ (taken from Gan (1990)).

McCool and Joyner-Motley (1998) presented a Bernoulli EWMA control chart for high-quality processes based on a transformation to an approximately normal variable. The normal approximation would be poor, if p is small and n is small to

Table 7.1 Comparison of the ARLs of the upper-sided Shewhart and modified EWMA control charts.

p	$\lambda = 1, h_l = 0, h_u = 33$	$\lambda = 0.580, h_l = 0, h_u = 28$	$\lambda = 0.14, h_l = 0, h_u = 23$
0.100	650.64	662.28	655.54
0.105	295.21	218.48	117.09
0.110	144.97	85.21	38.56
0.115	76.50	38.83	19.02
0.120	43.11	20.37	11.87
0.125	25.80	12.09	8.47
0.130	16.32	7.94	6.56
0.135	10.85	5.67	5.36
0.140	7.56	4.31	4.54
0.145	5.50	3.45	3.95
0.150	4.16	2.86	3.51
0.155	3.26	2.45	3.16
0.160	2.64	2.14	2.89
0.165	2.20	1.91	2.66
0.170	1.89	1.73	2.48
0.175	1.66	1.58	2.33
0.180	1.49	1.46	2.20
0.185	1.36	1.37	2.09
0.190	1.27	1.29	2.00
0.195	1.20	1.22	1.91
0.200	1.14	1.17	1.84

moderate sample sizes. Then, the improved square root transformation (ISRT) can be used to overcome this situation, which is applied in control chart for binomial observation. Sukparungsee (2014) proposed EWMA *p*-chart according to ISRT*p* and EWMA control charts. In using ISRT, the expected value of binomial random variable X is defined as

$$E(X) = \sqrt{p}$$

and the variance of X is

$$V(X) = \begin{cases} \left[\dfrac{1}{2}\sqrt{\dfrac{1-p}{n}} - \dfrac{1}{6} \cdot \dfrac{1-p}{n\sqrt{p}} \right]^2, & \text{for upper tail} \\[3ex] \left[\dfrac{1}{2}\sqrt{\dfrac{1-p}{n}} - \dfrac{3}{8} \cdot \dfrac{1-p}{n\sqrt{p}} \right]^2, & \text{for lower tail} \end{cases}$$

Therefore, the modified ISRT*p* EWMA control chart can present the control limits as the following:

$$\text{UCL}_{\text{ISRTEWMA}} = \sqrt{p} + L\sqrt{\frac{\lambda}{2-\lambda}\left[\frac{1}{2}\sqrt{\frac{1-p}{n}} - \frac{1}{6} \cdot \frac{1-p}{n\sqrt{p}} \right]}$$

$$\text{LCL}_{\text{ISRTEWMA}} = \sqrt{p} + L\sqrt{\frac{\lambda}{2-\lambda}\left[\frac{1}{2}\sqrt{\frac{1-p}{n}} - \frac{3}{8} \cdot \frac{1-p}{n\sqrt{p}} \right]} \tag{7.6}$$

where L is ISRT*p* EWMA chart coefficient.

Sukparungsee (2014) calculated run length performance of the ISRT*p* EWMA chart and compared with the run length performance of ISRT*p* chart. Table 7.2 gives the ARLs of both the control charts for $p = 0.05$, $n = 50$, $\lambda = 0.05$, and $\text{ARL}_0 = 370$, which are taken from Sukparungsee (2014).

Other works on the binomial EWMA version include Somerville, Montgomery, and Runger (2002), Yeh, Mcgrath, Sembower, and Shen (2008), Ross, Adams, Tasoulis, and Hand (2012), Spliid (2010), and Weiss and Atzmuller (2010).

7.1.2 Poisson EWMA (PEWMA) Chart

The Poisson distribution is frequently considered as a model for monitoring the number of nonconformities in a unit from a repetitive production process as discussed in Chapter 2. Borror, Champ, and Rigdon (1998) proposed the PEWMA chart to test average nonconformities per unit using the hypotheses:

$H_0 : \mu = \mu_0$ (Process is working at desired average nonconformities per unit)

$H_1 : \mu \neq \mu_0$ (Process is not working at desired average nonconformities per unit)

Table 7.2 Values of ARL of ISRTp and ISRTp EWMA control charts for $p = 0.05$, $n = 50$, $\lambda = 0.05$, and $ARL_0 = 370$.

δ	ISRTp, UCL = 0.399985	ISRTp EWMA, UCL = 3.03631
0	370.435	370.683
0.01	346.723	291.491
0.03	283.874	211.793
0.05	216.925	201.043
0.10	152.681	108.164
0.30	58.463	42.219
0.50	29.431	25.527
1.00	7.681	13.913
1.50	1.912	9.362
2.00	1.205	6.994

where μ_0 is the desired level of average nonconformities that should be as small as possible.

Let X_t, $t \geq 1$ be the number of nonconformities observed at time t, which are independent and identically distributed (i.i.d.) Poisson random variables with mean μ, where the in-control mean rate is $\mu = \mu_0$. The PEWMA statistic is defined as

$$Z_t = \lambda X_t + (1 - \lambda)Z_{t-1}, \quad t = 1, 2, \ldots \tag{7.7}$$

with $Z_0 = \mu_0$ and $\lambda(0 < \lambda < 1)$ be the smoothing constant. It is straightforward that $E(Z_t) = \mu_0$ and $V(Z_t) = \dfrac{\lambda\left[1 - (1-\lambda)^{2t}\right]\mu_0}{2 - \lambda}$. For large value of t, the variance is approximately $V(Z_t) \approx \frac{\lambda \mu_0}{2-\lambda} \approx V(Z_\infty)$.

The control limits are based on either the exact variance or the asymptotic variance. The time-varying control limits based on the exact control limits are

$$\left.\begin{aligned}
\mathrm{LCL}(t) &= \mu_0 - L_{\mathrm{L}}\sqrt{\frac{\lambda\left[1 - (1-\lambda)^{2t}\right]\mu_0}{2 - \lambda}} \\
\mathrm{CL} &= \mu_0 \\
\mathrm{UCL}(t) &= \mu_0 + L_{\mathrm{U}}\sqrt{\frac{\lambda\left[1 - (1-\lambda)^{2t}\right]\mu_0}{2 - \lambda}}
\end{aligned}\right\}, \tag{7.8}$$

The asymptotic control limits of PEWMA chart are

$$\left.\begin{aligned}
\text{LCL} &= \mu_0 - L_{\text{L}}\sqrt{\frac{\lambda\mu_0}{2-\lambda}} \\
\text{CL} &= \mu_0 \\
\text{UCL} &= \mu_0 + L_{\text{U}}\sqrt{\frac{\lambda\mu_0}{2-\lambda}}
\end{aligned}\right\}, \tag{7.9}$$

where $L = L_{\text{L}} = L_{\text{U}}$. The control chart multiplier L is usually set at 3 but can be adjusted to set the false alarm rate to a specified value, and it sometimes may be advantageous to construct asymmetric control limits. The PEWMA statistic defined in (7.7) is nonnegative. Thus, if the computed value of LCL is less than zero, then we set it equal to 0. The PEWMA chart raises an out-of-control signal when $Z_t < \text{LCL}(t)$ or LCL or $Z_t < \text{UCL}(t)$ or UCL.

Performance Evaluation Measure
The performance of an EWMA chart cannot be calculated analytically. However, the performance can be determined via simulation study and Markov chain approach. Borror et al. (1998) determined ARLs of the PEWMA chart for various values of the in-control mean using the Markov chain approach and using simulation.

Calculation of ARLs Using the Markov Chain Approach
The Markov chain approach can be used to approximate ARLs for shifts in the process mean. The interval $(h_{\text{L}}, h_{\text{U}})$ is divided into N subintervals. The jth subinterval is (L_j, U_j), where

$$L_j = h_{\text{L}} + \frac{(j-1)(h_{\text{U}} - h_{\text{L}})}{N}$$

and

$$U_j = h_{\text{L}} + \frac{j(h_{\text{U}} - h_{\text{L}})}{N}.$$

The midpoint, m_i, of the ith subinterval can be written as

$$m_i = h_{\text{L}} + 2\frac{(2i-1)(h_{\text{U}} - h_{\text{L}})}{N}.$$

The $(N + 1)$st state is absorbing and represents the out-of-control region above and below the control limits. This region is considered absorbing because the process is stopped when an out-of-control signal is raised. The ARL is thus the expected time to absorption of the Markov chain. The transition probability, P_{ij}, is the probability of moving from state i to state j in one step and is given by

$$P_{ij} = P(L_j < Z_t < U_j | Z_{t-1} = m_i).$$

This is the probability that Z_t is within the boundaries of state j, conditioned on Z_{t-1} being equal to the midpoint of state i. This transition probability can be written as

$$
\begin{aligned}
P_{ij} &= P\big(L_j < \lambda X_t + (1-\lambda)Z_{t-1} < U_j | Z_{t-1} = m_i\big)\\
&= P\big(L_j < \lambda X_t + (1-\lambda)m_i < U_j\big)\\
&= P\left(h_L + \frac{(j-1)(h_U - h_L)}{N} < \lambda X_t + (1-\lambda)\left(h_L + \frac{(2i-1)(h_U - h_L)}{2N}\right) < h_L + \frac{j(h_U - h_L)}{N}\right)\\
&= P\left(h_L + \frac{(h_U - h_L)}{N}(2(j-1) - (1-\lambda)(2i-1)) < X_t < h_L + \frac{(h_U - h_L)}{2N\lambda}(2j - (1-\lambda)(2i-1))\right)
\end{aligned}
$$

$$(7.10)$$

The random variable X_t, having a Poisson distribution, takes on only nonnegative integer values, although in general the left and right sides of Eq. (7.10) will not be integers. If, for example, the left and right sides of Eq. (7.10) are 4.6 and 6.1, respectively, then,

$$
\begin{aligned}
P_{ij} &= P(4.6 < X_t < 6.1)\\
&= P((X_t = 4.6) + P(X_t = 6.1))\\
&= \frac{\mu^5 \exp(-\mu)}{5!} + \frac{\mu^6 \exp(-\mu)}{6!}.
\end{aligned}
$$

The probability of transition to the out-of-control state can be determined in a similar manner, but it is not needed to obtain the ARLs. Let R_i be the ARL given that the process started in state i is given as

$$
Z_0 = m_i = P\left(h_L + \left(\frac{(2i-1)(h_U - h_L)}{2N}\right)\right).
$$

We define the vector R to be

$$
R = [R_1, R_2, ..., R_N]'.
$$

Let Q be the matrix obtained from the transition matrix P by deleting row $N+1$ and column $N+1$. In other words, Q is the transition matrix among the in-control states. Brook and Evans (1972) showed that the ARL vector R is the solution to the system

$$
(I - Q)R = 1,
$$

where I is the $N \times N$ identity matrix and where 1 is a $1 \times N$ column vector of ones. Thus, we can compute the ARLs for the PEWMA chart by solving the above linear system. The elements in the vector R are just the ARLs starting in the various states. The ARL given that $Z_0 = \mu_0$ is just the middle entry, that is, the $((N+1)/2)$th entry in the vector R. (We must choose N to be odd so that there is a unique middle value.)

The ARLs computed by this method are based on the assumption that the process shift occurs at the initial start-up of the system. Such ARLs are called *zero-state* or *initial-state* ARLs. Although not discussed in this paper, *steady-state* ARLs could be computed. The steady-state ARL assumes that the shift occurs long after the process has been operating in a state of statistical control. Lucas and Saccucci (1990) gave both the initial-state and steady-state ARLs for the EWMA based on the normal distribution. They found that the steady-state ARL is usually smaller than the initial-state ARL and that this difference increases as the smoothing constant, λ, decreases. For λ as small as 0.03, they found that the largest difference in the two ARLs is about 4%.

Borror et al. (1998) determined the ARLs of PEWMA by the Markov chain method using the IML procedure in SAS. The interval (h_L, h_U) was divided into $N = 101$ subintervals. The following tables give the ARL comparison of PEWMA chart, the Shewhart c-chart, and Gan's EWMA charts (taken from Borror et al. (1998)).

Table 7.3 shows that for nearly all shifts, the PEWMA charts with $\lambda = 0.57$ and $\lambda = 0.27$ have shorter ARLs than the Shewhart chart and any of the procedures proposed by Gan (1990).

Table 7.3 Comparison of the PEWMA chart, the Shewhart c-chart, and Gan's EWMA charts.

	$\lambda = 1$	$\lambda = 0.27$		$\lambda = 0.57$		$\lambda = 0.86$	
μ	Shewhart	CEWMA	PEWMA	CEWMA	PEWMA	CEWMA	PEWMA
6	3.5	3.7	2.2	2.6	2.1	2.3	2.3
8	10.0	5.0	2.7	3.9	2.7	4.5	4.5
10	34.2	7.9	3.4	7.8	4.2	12.0	12.1
12	131.6	18.9	4.8	26.2	8.6	43.6	43.9
14	553.7	138.5	8.5	172.7	28.4	202.5	203.0
16	2,426.2	5,566.2	24.2	2,002.1	164.9	1,119.0	1,114.6
18	4,876.2	89,655.2	197.2	14,341.5	1,486.3	4,391.9	4,380.6
20	1,218.6	1,234.5	1,233.4	1,384.5	1,383.7	1,291.4	1,291.8
22	266.1	78.8	85.2	160.9	168.7	238.8	238.5
24	76.0	18.1	17.7	34.4	35.5	61.2	61.2
26	27.5	8.4	7.8	12.1	12.3	21.1	21.1
28	12.2	5.4	4.8	6.1	6.1	9.4	9.4
30	6.4	4.0	3.5	3.9	3.8	5.1	5.1
35	2.2	2.5	2.2	2.0	2.0	2.0	2.0

Example 7.1   The data is taken from Borror et al. (1998), which representing the raw material of a process as shown in Table 7.4. The process is assumed to be operating at the mean level 4. Construct a PEWMA chart using $\lambda = 0.20$, and draw your conclusion about the state of process control.

Solution

For the given process, $\mu_0 = 4$ and $\lambda = 0.20$. The control charting constant is $L = 2.8275$. Thus, the asymptotic control limits are

$$\text{LCL} = \mu_0 - U_L \sqrt{\frac{\lambda \mu_0}{2 - \lambda}} = 4 - 2.8275 \sqrt{\frac{4 \times 0.20}{2 - 0.20}} = 2.115, \quad \text{CL} = \mu_0 = 4,$$

$$\text{UCL} = \mu_0 + U_L \sqrt{\frac{\lambda \mu_0}{2 - \lambda}} = 4 + 2.8275 \sqrt{\frac{4 \times 0.20}{2 - 0.20}} = 5.885.$$

Table 7.4   The example data of 40 batches of raw material.

t	X_t	Z_t	t	X_t	Z_t
1	5	4.2000	21	7	3.7910
2	3	3.9600	22	2	3.4328
3	4	3.9680	23	1	2.9462
4	0	3.1744	24	2	2.7570
5	2	2.9395	25	6	3.4056
6	9	4.1516	26	2	3.1245
7	2	3.7213	27	3	3.0996
8	2	3.3770	28	2	2.8797
9	4	3.5016	29	0	2.3037
10	1	3.0013	30	1	2.0430
11	2	2.8010	31	3	2.2344
12	6	3.4408	32	5	2.7875
13	5	3.7527	33	4	3.0300
14	1	3.2021	34	6	3.6240
15	7	3.9617	35	1	3.0992
16	3	3.7694	36	3	3.0794
17	2	3.4155	37	1	2.6635
18	0	2.7324	38	0	2.1308
19	4	2.9859	39	3	2.3046
20	3	2.9887	40	1	2.0437

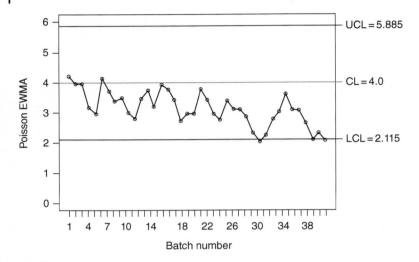

Figure 7.1   The Poisson EWMA chart for the raw material of a process.

The PEWMA statistic is calculated and given in the last column of Table 7.4. The PEWMA chart is constructed by plotting the EWMA statistic against the control limits in Figure 7.1. It is obvious that the process is out of control at the batch number 30, 37, and 40, and it is not working at average nonconformity level 4.

Testik, McCullough, and Borror (2006) studied the effect of estimated mean on the performance of PEWMA control charts. The control limits of a PEWMA chart with estimated parameter are

$$\left.\begin{array}{l} \widehat{LCL} = \widehat{\mu}_0 - L_L\sqrt{\dfrac{\lambda\widehat{\mu}_0}{2-\lambda}} \\[2mm] \widehat{CL} = \widehat{\mu}_0 \\[2mm] \widehat{UCL} = \widehat{\mu}_0 + L_U\sqrt{\dfrac{\lambda\widehat{\mu}_0}{2-\lambda}} \end{array}\right\}, \tag{7.11}$$

where $\widehat{\mu}_0 = \dfrac{1}{n}\sum_{i=1}^{n} X_i$ is maximum likelihood estimate of μ_0 based on n initial or reference sample size. The central limit theorem ensures that the sampling distribution of $\widehat{\mu}_0$ is approximately normal with mean μ_0 and variance $\dfrac{\mu_0}{n}$. The run length characteristics, such as ARL and standard deviation of run length (SDRL), were used to evaluate the performance of the PEWMA chart when parameter is estimated through again Markov chain approach discussed above. Table 7.5 gives the conditional performance of PEWMA chart when parameter is estimated.

Table 7.5 Conditional run length performance of the PEWMA chart with $\hat{\mu}_0 = 5$.

				75th		50th		25th	
n	*δ*	*λ*	*A*	**ARL**	**SDRL**	**ARL**	**SDRL**	**ARL**	**SDRL**
30	0	0.2	2.955	687.7	682.5	504.0	500.2	225.3	221.0
	1			13.4	9.6	10.2	6.8	8.2	5.1
	2			4.4	2.1	3.9	1.8	3.5	1.5
	3			2.7	1.0	2.5	0.9	2.4	0.9
	0	0.4	3.163	1035.6	1034.2	503.9	500.2	250.0	248.1
	1			17.1	14.9	12.4	10.3	9.5	7.5
	2			4.3	2.7	3.7	2.2	3.3	1.9
	3			2.4	1.2	2.2	1.1	2.0	1.0
100	0	0.2	2.955	675.0	670.8	504.0	500.2	327.3	323.2
	1			11.7	8.1	10.2	6.8	9.0	5.7
	2			4.1	1.9	3.9	1.8	3.4	1.6
	3			2.6	1.0	2.5	0.9	2.4	0.9
	0	0.4	3.163	752.9	751.4	503.9	500.2	340.4	338.6
	1			14.7	12.5	12.4	10.3	10.6	8.6
	2			4.0	2.5	3.7	2.2	3.5	2.0
	3			2.3	1.1	2.2	1.1	2.1	1.0
200	0	0.2	2.955	635.3	631.3	504.0	500.2	373.5	369.5
	1			11.3	7.7	10.2	6.8	9.3	6.0
	2			4.1	1.9	3.9	1.8	3.7	1.7
	3			2.6	1.0	2.5	0.9	2.5	0.9
	0	0.4	3.163	669.5	668.0	503.9	500.2	381.2	379.4
	1			13.9	11.8	12.4	10.3	11.1	9.0
	2			3.9	2.4	3.7	2.2	3.5	2.1
	3			2.2	1.1	2.2	1.1	2.1	1.0

They concluded that sample sizes greater than 300 for estimating the process mean would generally yield fairly close performance to the chart with known process mean. Shu, Jiang, and Wu (2012) extended the EWMA charts to Poisson processes with emphasis on quick detection of increases in Poisson rate for with and without normalizing transformation. Zero-state and steady-state average length was calculated using a Markov chain model, and the comparison results indicated that the EWMA chart based on normalized data was nearly optimal. Abujiya,

Farouk, Lee, and Mohamad (2013) proposed an EWMA chart with fast initial response (FIR) using supplementary run rules (2/2) scheme for monitoring Poisson observation. Sukparungsee (2014) presented the combined ISRT and EWMA control charts together based on binomial observations, namely, ISRTp EWMA chart for which the performance of this modified control chart is better than ISRTp chart by measuring the ARL.

A lot of recent articles focus on the PEWMA and cumulative sum (CUSUM) charts with time-varying sample sizes, see, e.g. Dong, Hedayat, and Sinha (2009), Ryan and Woodall (2010), and Mei, Han, and Tsui (2011).

7.1.3 Other EWMA Charts

A lot of work has been done in the literature on the EWMA control charts for geometric and negative binomial distribution, but a few articles considered these charts as an alternative to monitor over-dispersion in count data sets.

Geometric EWMA Chart

Sun and Zhang (2000) considered developing an EWMA chart with the asymptotic control limits for monitoring geometric random variables. If X_t, $t = 1, 2, ...$, are i.i.d. geometric random variables with probability p, then the EWMA statistics are defined as

$$Z_t = \lambda X_t + (1 - \lambda)Z_{t-1}, \quad t = 1, 2, ... \tag{7.12}$$

with $Z_0 = \dfrac{p_0}{1 - p_0}$ is the historical average of the data and $\lambda(0 < \lambda < 1)$ be the smoothing constant. Rearranging the EWMA statistic (7.12), we have

$$Z_t = \lambda X_t + (1 - \lambda)Z_{t-1} = Z_{t-1} + \lambda(X_t - Z_{t-1}), \quad t = 1, 2, ... \tag{7.13}$$

Hence, the statistic Z_t can be viewed as the forecast for the next value X_{t+1}; it is equal to the present predicted value plus λ times the present observed error of prediction.

The control chart raises an out-of-control signal when $Z_t < \text{LCL}$ or $Z_t > \text{UCL}$, where the asymptotic limits are

$$\left.\begin{aligned}
\text{LCL} &= \frac{p_0}{1 - p_0} - L_L\sqrt{\frac{\lambda}{2 - \lambda}\frac{p_0}{(1 - p_0)^{2t}}} \\
\text{CL} &= \frac{p_0}{1 - p_0} \\
\text{UCL} &= \frac{p_0}{1 - p_0} + L_U\sqrt{\frac{\lambda}{2 - \lambda}\frac{p_0}{(1 - p_0)^{2t}}}
\end{aligned}\right\} \tag{7.14}$$

where L_L and L_U are control charting constants for two-sided geometric EWMA chart. Sun and Zhang (2000) designed the geometric EWMA chart based on an acceptable in-control average number of nonconformity (ANNC), which is defined as

ANNC = average of the expected number of items inspected until a signal is produced.

where ANNC(0) refers to in-control ANNC at $p = p_0$ and ANNC(1) means the ANNC when proportion of nonconformities shifted from p_0 to p_1. An optimal chart is defined here as the chart with a fixed in-control ANNC(0) that has the smallest ANNC for a given shift.

Sun and Zhang (2000) determined a large range of L_L and L_U values for a given fixed value of p_0 to obtain a desired in-control ARL value for their chart using algorithm. Table 7.6 gives the one-sided L_L and L_U values for geometric EWMA chart at ANNC 150 and 200 (taken from Sun and Zhang (2000)).

The control limits defined in Eq. (7.13) are sensitive to over- or under-dispersed data as discussed by many authors for c-chart due to constraint of equi-dispersion for Poisson distribution.

The performance of the geometric EWMA chart can be measured either by Markov chain or by Monte Carlo simulation procedure. Sun and Zhang (2000) found the zero-state average number of signal (ANOS) as performance measure using simulation. An EWMA chart for monitoring a negative binomial random variable was proposed by Kotani, Kusukawa, and Ohta (2005).

If X_t, in Eq. (7.12), is the cumulative count of item inspected until observing r nonconforming ones, assumed to obey the negative binomial distribution with parameters r and p. When the process is in control, the mean and variance of a negative binomial process are

Table 7.6 The one-sided control charting constant for geometric EWMA chart.

λ	ANNC = 150		ANNC = 200		λ	ANNC = 150		ANNC = 200	
	L_l	L_u	L_l	L_u		L_l	L_u	L_l	L_u
0.05	1.4492	1.751	1.5777	1.979	0.35	1.5177	3.248	1.5565	3.492
0.10	1.6220	2.234	1.7200	2.457	0.40	1.4710	3.369	1.5060	3.626
0.15	1.6425	2.543	1.7229	2.775	0.50	1.3840	3.569	1.4070	3.827
0.20	1.6300	2.767	1.6900	2.998	0.75	1.1723	3.867	1.1846	4.133
0.25	1.5971	2.938	1.6531	3.177	0.90	1.0560	3.968	1.0620	4.220
0.30	1.5610	2.938	1.6050	3.344	1.00	0.9937	3.997	0.9951	4.233

$$E(X_t) = \frac{r}{p_0} \text{ and } V(X_t) = \frac{r(1-p_0)}{p_0^2}, \text{respectively.} \qquad (7.15)$$

The control limits for the negative binomial EWMA chart are

$$\left. \begin{array}{l} \text{LCL} = \dfrac{r}{p_0} - L\sqrt{\dfrac{\lambda}{2-\lambda}\dfrac{r(1-p_0)}{p_0^2}} \\[3mm] \text{CL} = \dfrac{r}{p_0} \\[3mm] \text{UCL} = \dfrac{r}{p_0} + L\sqrt{\dfrac{\lambda}{2-\lambda}\dfrac{r(1-p_0)}{p_0^2}} \end{array} \right\} \qquad (7.16)$$

where L is width of the control limits of negative binomial EWMA chart and can be determined for fixed false alarm probability.

Kotani et al. (2005) studied the performance of the negative binomial EWMA chart using ANOS as performance measure. Assuming the in-control proportion of nonconformities $p = p_0$ against the shifted proportion of nonconformities $p_1 = \kappa p_0$, the ANOS can be defined as

$$\text{ANOS} = \frac{1}{P_L + P_U}, \qquad (7.17)$$

where $P_U = \left(1 - \sum_{i=r}^{\text{UCL}} \binom{i-1}{r-1} p_1^r (1-p_1)^{i-r}\right)^2$ and $P_L = \left(\sum_{i=r}^{\text{LCL}} \binom{i-1}{r-1} p_1^r (1-p_1)^{i-r}\right)^2$. Table 7.7 gives the ANOS values of negative binomial EWMA chart determined by Kotani et al. (2005) using Markov chain approach.

Conway–Maxwell–Poisson (COM–Poisson) EWMA Chart

Saghir and Lin (2014b) designed a generalized EWMA chart based on COM–Poisson distribution for monitoring count data. This control chart is flexible for under- or over-dispersed data that exist in many count data sets.

The mean and variance of COM–Poisson random variable X having probability density function (p.d.f.) defined in (2.21) are

$$\left. \begin{array}{l} E(X) = \mu\dfrac{\partial \log\left(Z(\mu,\upsilon)\right)}{\partial\mu} \approx \mu^{1/\upsilon} - \dfrac{\upsilon-1}{2\upsilon} \\[3mm] \text{Var}(X) = \dfrac{\partial E(X)}{\partial\mu} \approx \dfrac{1}{\upsilon}\mu^{1/\upsilon} \end{array} \right\}, \qquad (7.18)$$

where approximation particularly holds for $\mu > 10^\upsilon$.

Table 7.7 ANOS values of negative binomial EWMA chart for $p_0 = 0.001$, $r = 2$ and 5, and $\alpha_0 = 0.0027$.

κ	$r = 2$ (λ, L)			$r = 5$ (λ, L)		
	(0.06, 2.563)	(0.07, 2.626)	(0.08, 2.684)	(0.06, 2.563)	(0.07, 2.626)	(0.08, 2.684)
0.5	8	7	7	5	4	4
0.6	12	11	11	7	7	6
0.7	20	19	19	11	11	11
0.8	39	39	39	22	22	22
0.9	104	104	104	66	66	66
1.0	0.2694	0.2698	0.2699	0.2694	0.2697	0.2697
1.1	366	460	584	140	161	185
1.2	138	175	231	48	52	57
1.3	70	84	104	27	28	30
1.4	45	51	60	19	20	20
1.5	34	37	41	15	15	16

In the case of unknown parameters μ and ν, we can estimate the parameters via maximum likelihood estimation (MLEs). The MLEs of the parameters can be obtained as

$$L\left(\mu, \nu \middle| X\right) = \prod_{i=1}^{n} \frac{\mu^{x_i}}{x! Z(\mu, \nu)} \tag{7.19}$$

Equation (7.19) could not be solved analytically; therefore, an iterative method such as the Newton–Raphson method can be used, or *compoisson* package in *R* can be used.

Let X_t, $t = 1, 2, \ldots$ are i.i.d. COM–Poisson random variables. When the process is in control, the process mean is $\mu_0 \dfrac{\partial \log\left(Z(\mu_0, \nu)\right)}{\partial \mu_0}$. Here we are interested in monitoring changes in the average number of nonconformities (μ) for the fixed level of data dispersion (ν). Therefore, the EWMA statistics are defined as

$$W_t = \lambda X_t + (1 - \lambda) W_{t-1}, \quad t = 1, 2, \ldots \tag{7.20}$$

with $W_0 = \mu_0 \dfrac{\partial \log (Z(\mu_0, v)}{\partial \mu_0}$ average value of COM–Poisson random variable and $\lambda (0 < \lambda < 1)$ be the smoothing constant. The control limits of the COM–Poisson EWMA chart based on COM–Poisson distribution are defined as

$$
\left.
\begin{aligned}
\text{LCL} &= \mu_0 \frac{\partial \log (Z(\mu_0, v))}{\partial \mu_0} - L_{\text{L}} \sqrt{\frac{\left|\dfrac{\partial E(X)}{\partial \mu_0}\right| \lambda \left[1 - (1 - \lambda)^{2t}\right]}{2 - \lambda}} \\
\text{CL} &= \mu_0 \frac{\partial \log (Z(\mu_0, v))}{\partial \mu_0} \\
\text{UCL} &= \mu_0 \frac{\partial \log (Z(\mu_0, v))}{\partial \mu_0} + L_{\text{U}} \sqrt{\frac{\left|\dfrac{\partial E(X)}{\partial \mu_0}\right| \lambda \left[1 - (1 - \lambda)^{2t}\right]}{2 - \lambda}}
\end{aligned}
\right\}
\tag{7.21}
$$

For large values of t, the asymptotic limits are

$$
\left.
\begin{aligned}
\text{LCL} &= \mu_0 \frac{\partial \log (Z(\mu_0, v))}{\partial \mu_0} - L_{\text{L}} \sqrt{\frac{\lambda \left|\dfrac{\partial E(X)}{\partial \mu_0}\right|}{2 - \lambda}} \\
\text{CL} &= \mu_0 \frac{\partial \log (Z(\mu_0, v))}{\partial \mu_0} \\
\text{UCL} &= \mu_0 \frac{\partial \log (Z(\mu_0, v))}{\partial \mu_0} + L_{\text{U}} \sqrt{\frac{\lambda \left|\dfrac{\partial E(X)}{\partial \mu_0}\right|}{2 - \lambda}}
\end{aligned}
\right\}
\tag{7.22}
$$

The control chart multiplier $L = L_{\text{L}} = L_{\text{U}}$ can be selected according to the pre-specified probability of false alarm after choosing the smoothing constant λ and parameters of COM–Poisson distribution.

The strength of the COM–Poisson EWMA chart is its flexibility and ability to generalize PEWMA chart developed via the Poisson assumption, the Bernoulli EWMA chart from the Bernoulli distribution, and geometric EWMA chart from the geometric distribution. The control limits of the COM–Poisson EWMA coincide with usual PEWMA chart limits for $v = 1$, Bernoulli EWMA chart for $(v \to \infty$ and $p = \mu_0/(1 + \mu_0))$, and geometric EWMA chart for $(v = 0, \mu_0 < 1)$. This is very helpful when the underlying count distribution is unknown. Thus, in-control data can be modeled well via the COM–Poisson EWMA control chart for good performance. For over-dispersion data approximation can easily be satisfied as for $v < 1$, $\mu > 10^v$ easily chosen in practice; therefore, approximated mean and variance values could be used to make it more simples for calculation.

The values of L_{L} and L_{U} are often chosen to be equal ($L_{\text{L}} = L_{\text{U}} = L$). For some choices of parameters, the symmetric limits produce negative LCL, then we set

Table 7.8 Charting constant L for COM–Poisson EWMA chart at $ARL_0=200$.

					λ				
(μ, ν)	0.01	0.05	0.08	0.10	0.15	0.20	0.25	0.50	0.75
(1, 0.5)	1.482	2.167	2.347	2.422	2.565	2.696	2.818	3.235	3.462
(1, 1.0)	1.484	2.183	2.366	2.432	2.562	2.653	2.746	3.141	3.252
(1, 10)	1.520	2.148	2.284	2.337	2.377	2.453	2.514	2.725	2.842
(2, 0.5)	1.487	2.172	2.335	2.407	2.525	2.611	2.668	2.920	3.065
(2, 1.0)	1.490	2.192	2.350	2.418	2.526	2.621	2.689	2.955	3.095
(2, 10)	1.495	2.205	2.313	2.364	2.433	2.500	2.542	2.805	2.885
(3, 1.0)	1.486	2.186	2.344	2.410	2.533	2.608	2.664	2.865	2.984

LCL $= 0$, and the COM–Poisson EWMA chart will not signal at a lower control limit that is less than or equal to 0. After setting the control limits for the proposed chart, the EWMA statistic given in Eq. (7.20) is plotted against t. For an in-control process, most of the W_t's should lie inside the control limits, whereas an out-of-control process is signaled by one or more of the W_t's lying outside the control limits.

Saghir and Lin (2014b) determined the wide range of control charting constants using Monte Carlo simulation. Table 7.8 gives the charting constant value for some specific parameters and false alarm rate.

Saghir and Lin (2014b) studied the performance of COM–Poisson EWMA chart via Monte Carlo simulation. The ARL is used as performance measure. Table 7.9 gives the simulated ARL value of COM–Poisson EWMA chart.

Example 7.2 Consider the following data concerning the nonconformities observed in 26 samples of circuit boards (see Montgomery (2013, p. 173)). Apply COM–Poisson EWMA chart to identify the state of the process (Table 7.10).

Solution

The optimization scheme gives the COM–Poisson MLEs, $\hat{\mu} = 2.871$, and $\hat{\nu} = 0.365$ (using R). This indicates that data has over-dispersion and the Poisson distribution is not a suitable model to fit this data due to its variance constraint. Consider $\lambda = 0.15$ and $L = 3.001$, the control limits of COM–Poisson EWMA chart are

$$\text{LCL} = \mu_0 \frac{\partial \log (Z(\mu_0, \upsilon))}{\partial \mu_0} - L_1 \sqrt{\frac{\lambda \dfrac{\partial E(X)}{\partial \mu_0}}{2 - \lambda}} = \mu^{1/\upsilon} - \frac{\upsilon - 1}{2\upsilon} - 3.001 \sqrt{\frac{\lambda \dfrac{1}{\upsilon} \mu^{1/\upsilon}}{2 - \lambda}} = 12.878$$

Table 7.9 ARL for the COM–Poisson distribution at $ARL_0 = 200$ and $\nu = 0.5$.

μ	$\lambda = 0.05$	$\lambda = 0.20$	$\lambda = 0.50$	$\lambda = 1.00$
3.50	1.0	10.8	51.0	554.4
3.70	1.5	40.7	144.1	1066.4
3.75	6.5	57.0	195.1	1190.9
3.80	20.9	82.9	244.1	687.6
3.85	47.1	121.7	294.6	561.7
3.90	88.9	174.9	281.8	414.7
3.95	151.7	215.9	248.7	289.5
4.00	**200.5**	**201.4**	**199.8**	**211.3**
4.05	151.2	157.1	162.3	185.8
4.10	83.5	110.2	113.7	132.5
4.15	41.8	72.4	78.8	98.6
4.20	18.1	48.1	55.4	84.3
4.25	4.5	33.8	44.1	72.8
4.30	1.2	23.5	34.6	60.4
4.50	1.0	5.9	12.7	33.9

Bold values represents in-control ARL.

Table 7.10 Observed nonconformities in 26 samples of circuit boards.

Circuit board no.	Observed nonconformities	Circuit board no.	Observed nonconformities
1	19.17	14	20.14
2	19.89	15	18.62
3	19.31	16	18.37
4	18.21	17	17.57
5	17.73	18	18.23
6	15.82	19	18.2
7	17.64	20	21.32
8	18.0	21	22.62
9	19.95	22	22.83
10	20.7	23	21.8
11	20.6	24	21.38
12	21.11	25	20.72
13	20.34	26	19.86

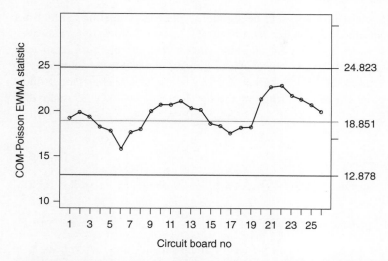

Figure 7.2 The COM–Poisson EWMA chart for the number of defective in circuit board.

$$CL = \mu_0 \frac{\partial \log\left(Z(\mu_0, v)\right)}{\partial \mu_0} = \mu^{1/v} - \frac{v-1}{2v} = 18.851$$

$$LCL = \mu_0 \frac{\partial \log\left(Z(\mu_0, v)\right)}{\partial \mu_0} + L_u \sqrt{\frac{\lambda \frac{\partial E(X)}{\partial \mu_0}}{2 - \lambda}} = \mu^{1/v} + \frac{v-1}{2v} + 3.001 \sqrt{\frac{\lambda \frac{1}{v} \mu^{1/v}}{2 - \lambda}} = 24.823$$

and 24.823 with CL at 18.851 (see Figure 7.2).

The figure shows that all the number of defective lies within the control limits of the EWMA chart and declared that the process is in control at or working the desired level of average defectives. These initial limits are also called *trial limits*.

7.2 CUSUM Control Charts for Attributes

Similar to the EWMA chart, a CUSUM chart dominates the Shewhart chart in detecting small to moderate-sized shifts in the parameter of interest in terms of statistical performance. According to Szarka and Woodall (2011), the choice between the use of a CUSUM chart and an EWMA chart depends largely on the personnel preference of the user. However, CUSUM approach is preferable due to its optimality properties. For a comprehensive overview of CUSUM charts for numerous distributions, see Hawkins and Olwell (1998) and Koshti (2011).

A CUSUM chart is designed to be optimal against all competing methods for detecting a specified change in a parameter of interest θ from θ_0 to θ_1. However, θ_1 should be chosen such that the chart has good statistical range performance for a wide of shifts. In attribute control charts, only an increase in θ is of interest to detect, so a one-sided CUSUM chart is generally used. CUSUM charts for attribute data were proposed at early stage in the development of SPC and referred in Page (1962).

7.2.1 Binomial CUSUM Chart

Gan (1993) proposed a CUSUM chart for binomial counts based on the CUSUM statistics

$$S_t = \text{Max}(0, S_{t-1} + X_t - k), \quad t = 1, 2, \ldots \tag{7.23}$$

where X_t, $t = 1$, 2, ..., is a sequence of independent binomial counts with parameters n and p, and k is a suitable chosen constant, and $S_0 = u$, $0 \leq u \leq H$. This CUSUM chart is intended for detecting upward shifts in p, and it issues an out-of-control signal at the first t for which $S_t > H$. He also provided a way for determining the optimal values of k and the charts limit H corresponding to a particular shift for a fixed in-control ARL and p_0 spanning from 0.05 to 0.20.

There is a close relationship between the CUSUM chart and the sequential probability ratio test (SPRT). Let the null hypothesis be $H_0 : p = p_0$ and the alternative hypothesis be $H_1 : p = p_1$, where $p > p_1$. The SPRT employs the likelihood ratio sequentially as

$$\lambda_t = \frac{\prod_{i=1}^{t} \text{Pr}(X = x_i; p_0)}{\prod_{i=1}^{t} \text{Pr}(X = x_i; p_1)} \quad t = 1, 2, \ldots \tag{7.24}$$

An equivalent test involving the natural logarithm of λ_t is often used. For binomial counts, $\ln(\lambda_t)$ can be expressed as

$$\ln(\lambda_t) = \sum_{i=1}^{t}(X_i - k_{\text{SPRT}}) = \ln(\lambda_{t-1}) + X_t - k_{\text{SPRT}}$$

where the reference value of the SPRT, k_{SPRT}, is given by

$$k_{\text{SPRT}} = \frac{n \ln((1 - p_0)/(1 - p_1))}{\ln((1 - p_0)/(1 - p_1)) - \ln(p_0/p_1)} \tag{7.25}$$

The test continues as long as $A_0 < \ln(\lambda_t) < A_1$, for some fixed A_0 and A_1; accept H_0, if $\ln(\lambda_t) < A_0$, and reject H_0, if $\ln(\lambda_t) > A_1$. Unlike the CUSUM chart, the SPRT does not set $\ln(\lambda_t)$ to zero when the sum is negative. When $\ln(\lambda_t) < A_0$, this will lead to an acceptance of H_0. In other words, the CUSUM chart does not terminate when H_0 is true (the process is in control) for a fixed type I error.

A typical comparison is displayed in Table 7.1. The in-control mean is assumed to be $p = 0.05$ and the ARL profile of seven CUSUM charts using k_{SPRT}, with $p_1 = 0.053$, 0.056, 0.059, 0.062, 0.065, 0.066, and 0.068 are displayed in Table 7.11. The chart limits of these charts are selected such that they have an in-control ARL of approximately 240. Due to the discrete nature of the binomial distribution, sometimes it is not possible to find the chart limit such that the ARL of the associated CUSUM chart agrees exactly with the specified in-control ARL, which is 240 in this case. These CUSUM charts are found to be optimal at the respective values of p_1. Similar comparisons using other values of n, p_0, p_1, and in-control ARL all show that the reference value of the SPRT is in fact the optimal value of k for the CUSUM chart.

Reynolds and Stoumbos (2000) introduced a CUSUM chart based on applying a SPRT at each sampling point for Bernoulli data for interval sampling. They compared their chart with the p-chart and binomial CUSUM chart. The SPRT chart is flexible because of the use of sequential testing, and it gives mostly superior results over the two competing methods using the steady-state average time to signal metric when considering an increase from $p_0 = 0.01$ for a wide range of shifts. Wu, Jiao, and Liu (2008) considered a binomial CUSUM chart to detect large shift in p and designed CUSUM statistic altered to improve the performance of these shifts. This chart compared favorably with the traditional binomial CUSUM chart for large values of γ for two cases. However, the traditional binomial CUSUM chart would be expected to perform best overall because of its advantage in detecting smaller shifts.

Reynolds and Stoumbos (2000) studied the binomial CUSUM chart and a special case of this chart, where $n = 1$, which was termed the Bernoulli CUSUM chart. A primary advantage of this method, providing calculations are automated, is that a CUSUM statistic is calculated after each item is inspected. The upper-sided Bernoulli CUSUM statistics, B_t, $t = 1, 2, \ldots$, are

$$B_t = \text{Max}(0, B_{t-1} + X_t - k_B), \quad t = 1, 2, \ldots \tag{7.26}$$

where $B_0 = 0$ and $k_B = 1/k_G$. A signal is raised if $B_t \geq h_B$. They evaluated the properties of the p-chart, the binomial CUSUM, and Bernoulli CUSUM charts using upper-sided control limits, focusing on an increase in p, where $p_0 = 0.01$ and different values of n were considered for the p-chart ($n = 51$, 100, 158) and the binomial CUSUM chart ($n = 51$, 100). Using the zero-state ANOS to evaluate performance, the Bernoulli CUSUM chart was shown to be the uniformly best method for detecting all shifts in p. The optimal performance for a binomial CUSUM chart when $n = 1$ (the Bernoulli CUSUM chart) was also shown by

Table 7.11 Comparison of ARLs of control charts with respect to both in-control and out-of-control process fraction of nonconforming items produced.

p	$n=100$ $k=5.13$ $H=23.75$ $p_{opt}=0.053$	$n=100$ $k=5.29$ $H=18.29$ $p_{opt}=0.056$	$n=100$ $k=5.44$ $H=15.00$ $p_{opt}=0.059$	$n=100$ $k=5.57$ $H=13.14$ $p_{opt}=0.062$	$n=100$ $k=5.71$ $H=11.57$ $p_{opt}=0.065$	$n=100$ $k=5.78$ $H=11.00$ $p_{opt}=0.066$	$n=100$ $k=5.88$ $H=10.25$ $p_{opt}=0.068$
0.050	240.5	240.1	239.8	240.4	240.3	240.4	240.9
0.051	157.5	161.2	165.6	169.5	173.1	174.6	177.4
0.052	110.9	113.9	118.7	123.1	127.7	129.7	133.1
0.053	82.9	84.4	88.2	92.1	96.5	98.4	101.9
0.054	65.2	65.1	67.8	70.9	74.6	76.3	79.3
0.055	53.2	52.1	53.7	56.0	59.0	60.4	62.9
0.056	44.7	43.0	43.7	45.3	47.7	48.8	50.8
0.057	38.5	36.3	36.5	37.5	39.3	40.2	41.8
0.058	33.7	31.3	31.0	31.7	32.9	33.6	34.9
0.059	30.0	27.5	26.9	27.2	28.1	28.6	29.6
0.060	27.0	24.4	23.6	23.7	24.3	24.7	25.4
0.061	24.5	22.0	21.0	21.0	21.3	21.6	22.2
0.062	22.5	19.9	18.9	18.7	18.9	19.1	19.5
0.063	20.8	18.3	17.2	16.9	16.9	17.0	17.3
0.064	19.3	16.8	15.7	15.4	15.3	15.4	15.6
0.065	18.0	15.6	14.5	14.1	13.9	14.0	14.1

0.066	16.9	14.6	13.5	13.0	12.8	12.8	12.9
0.067	15.9	13.7	12.5	12.1	11.8	11.8	11.8
0.068	15.0	12.9	11.7	11.3	11.0	10.9	10.9
0.069	14.2	12.1	11.0	10.6	10.2	10.2	10.1
0.070	13.5	11.5	10.4	9.6	9.6	9.5	9.4
0.080	9.0	7.6	6.7	6.3	5.9	5.8	5.6
0.090	6.9	5.7	5.0	4.6	4.3	4.3	4.0
0.100	5.6	4.6	4.0	3.7	3.4	3.3	3.2
0.200	2.1	1.8	1.6	1.4	1.3	1.2	1.2

Bourke (1999) using zero-state and steady-state analyses with the random-shift model. Megahed, Kensler, Bedair, and Woodall (2011) considered an ANOS property of the two-sided Bernoulli CUSUM chart. Reynolds (2013) considered the Bernoulli CUSUM chart for detecting a decrease in the proportion p of nonconforming items when a continuous stream of Bernoulli observation from the process is available. In this article a solution to the equations is used in defining the Markov chain so that properties of the chart can be obtained without explicitly using the transition matrix was provided.

Reynolds and Stoumbos (2000) developed the binomial and Bernoulli CUSUM charts when 100% inspection was not necessarily present. These CUSUM charts were also compared with a p-chart for detecting an increase in p. Joner, Woodall, and Reynolds (2008) examined a scan statistic and compared this with the Bernoulli CUSUM chart. The Bernoulli scan method is based on a moving window of past data. If the number of nonconforming items in the most recent window of items exceeds a specified value, a signal is given. Considering a large range of shifts, the Bernoulli CUSUM statistic had better performance for detecting an increase in p on the basis of the random-shift steady state average number of observations to signal (SSANOS) metric. Sego, Woodall, and Reynolds (2008) considered three different "sets" methods versus the Bernoulli CUSUM chart. The sets methods are based on runs rules-type procedures based on the successive counts of conforming items between nonconforming items. The results of their study showed that the Bernoulli CUSUM chart was inferior for zero-state analysis (even when a head-start feature was added), but was uniformly best when the random-shift steady-state analysis was considered.

Mousavi and Reynolds (2009) developed a Markov binary CUSUM chart for monitoring correlated Bernoulli data. It was assumed that a positive correlation exists with a first-order dependence, represented by a two-state Markov model. As the autocorrelation coefficient ρ becomes larger in magnitude, the in-control ANOS values based on the Bernoulli CUSUM charts become quite small. Hence, the standard Bernoulli CUSUM methods become unreliable, even for smaller values of ρ, because the actual average number of observations to signal at in-control (ANOS_0) values are much lower than for the special case with independent observations, where $\rho = 0$. Lee, Wang, Xu, Schuh, and Woodall (2013) studied the effect of parameter estimation on upper-sided Bernoulli CUSUM charts. The expected value of the ANOS and the standard deviation of the ANOS were used as performance measures in that study. It was found that the effects of parameter estimation on the Bernoulli CUSUM chart are more significant than those on the Shewhart-type geometric chart. A large sample in phase-I is required for the Bernoulli CUSUM chart to produce the desired in-control ANOS.

7.2.2 Poisson CUSUM Chart

Let $X_i, t = 1, 2, \ldots$ be a sequence drawn from continuous Poisson process. To detect an increase in counts, the CUSUM statistic is

$$S_i = \text{Max}(0, S_{i-1} + X_i - k), \quad i = 1, 2, \ldots \tag{7.27}$$

where $\text{Max}(a, b)$ is the maximum of a and b. A standard CUSUM will have starting value $S_0 = 0$, whereas an FIR CUSUM will use a positive starting value. Lucas (1985) recommend a head start (S_0) value approximately equal to $h/2$ and provided values of h and k for various parametric values of Poisson parameter μ.

In the design of a Poisson CUSUM chart, the reference value k is determined to distinguish between an acceptable mean value μ_0 and a detectable mean value μ_d (a shifted average level) that the CUSUM is to detect quickly. The reference value k is given by

$$k = \frac{\mu_d - \mu_0}{\ln(\mu_d) - \ln(\mu_0)} \tag{7.28}$$

Generally, μ_0 is chosen to be the current system performance, i.e. μ_0 if the in-control mean is known or $\widehat{\mu_0}$ if it is estimated. The μ_d value represents the out-of-control mean for which the CUSUM is being tuned to detect quickly. Note that the reference value is related to the magnitude of the shift to be detected and it also serves as a threshold value, which allows CUSUM to accumulate if the counts are above this value. This reference value is the same as the reference value for an SPRT. However, unlike the normal distribution case where the reference value is midway between the in-control and out-of-control mean values, for the Poisson case, k lies between μ_0 and μ_d, lying a bit closer to μ_0. The design and implementation of CUSUM charts to detect both increases and decreases in the count level along with the FIR feature and the robust CUSUM were discussed in White, Keats, and Stanley (1997). These enhancements speed up the detection of changes in the count level and guard against the effects of atypical or outlier observations.

White et al. (1997) suggested the procedures for constructing a Poisson CUSUM chart to monitor the detection of an increase in the number of nonconformities. The ARLs of the proposed Poisson CUSUM chart for in-control and out-of-control situations were calculated using Markov chain approach and compared with the corresponding ARLs of the c-chart. The results of study show that in most cases, the Poisson CUSUM is able to provide superior protection against a shift while maintaining a lower false alarm rate. For illustration, the ARL curves for Poisson CUSUM chart versus c-chart are constructed in Figure 7.3a and b.

Ryan and Woodall (2010) recommended a CUSUM and an EWMA control charts to monitor a Process with Poisson count data when sample size varies. The ability of these charts to detect increases in the Poisson rate was evaluated

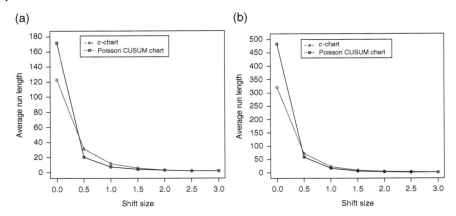

Figure 7.3 ARL curves based on Poisson CUSUM chart and c-chart for (a) μ_0 = 4 and μ_d = 6.0 and (b) μ_0 = 12 and μ_d = 24.0.

by calculating the steady-state ARL performance. Results indicated that the CUSUM chart is best at monitoring Poisson count data at the out-of-control shift when the sample size varies randomly. In addition, a EWMA method was proposed that has good steady-state ARL performance. Testik et al. (2006) analyzed the effect of estimated process mean on the conditional and marginal performance of the Poisson CUSUM chart. The Markov chain approach was used for calculating the aspects of the run length distribution. The run length characteristics such as ARL, SDRL, and percentiles of run length distributions were calculated. The effect of estimation on the in-control ARL performance was shown to be significant, and also sample size recommendations were made.

Han, Tsui, Ariyajunya, and Kim (2010) compared the performance of the three detection methods: temporal scan statistic, CUSUM, and EWMA when the observations follow the Poisson distribution in health care. The results of simulation study showed that the Poisson CUSUM and EWMA charts generally outperformed the Poisson scan statistic methods. The CUSUM charts were superior in dealing with a large shift with a later change in time in comparison with EWMA charts. However, the EWMA charts outperformed the CUSUM charts in situations with a small shift and an early change in time.

Perry and Pignatiello (2011) studied the effect of changes in the design of the control chart on the performances of the change point estimators of CUSUM and EWMA Poisson charts. They compared root mean square error performances of the change point estimators offered by the Poisson CUSUM and EWMA control charts relative to that achieved by a maximum likelihood estimator for the process change point. The results indicated that the relative performance achieved by each change point estimator is a function of the corresponding control chart design. Relative mean index plots are provided to enable users of these control charts to choose a control chart design and change point estimator combination that yield robust change point estimation performance across a range of potential change magnitudes. A two-sided CUSUM chart for autocorrelated count processes using a Poisson integer-valued autoregressive model of order 1, namely, Poisson INAR (1) model was developed by Yontay, Weiss, Testik, and Bayindir (2013). A trivariate Markov chain approach was developed for exact evaluation of the ARL performance of the chart in addition to a computationally efficient approximation based on bivariate Markov chains. The design of the chart for an ARL-unbiased performance and the analyses of the out-of-control performances were discussed.

7.2.3 Geometric CUSUM Chart

Bourke (2001) first proposed a CUSUM chart based on the geometric random variables. The upper-sided CUSUM statistics for a geometric random variable are as follows:

$$G_t = \text{Max}(0, G_{t-1} - Y_t + k_G), \quad t = 1, 2, \dots \tag{7.29}$$

where $G_0 = 0$ and k_G can often be conveniently specified to be an integer. The expression of k_G based on log-likelihood ratio for optimally detecting p_0 to p_1 is defined as

$$k_G = \frac{\ln\left(\dfrac{p_1(1-p_0)}{p_0(1-p_1)}\right)}{\ln\left(\dfrac{1-p_0}{1-p_1}\right)}. \tag{7.30}$$

A signal for this chart is raised if $G_t \geq h_G$, where h_G is known as the decision interval or decision limit. The upper or lower or two-sided geometric CUSUM can be designed as per requirement of detecting shifts in either upper or lower or both directions.

Bourke (1999) developed the geometric CUSUM chart with sampling inspection for monitoring fraction defective. A large-scale analysis of the ANOS values for this chart was presented for p_0 with different charting parameters. Sun and Zhang (2000) designed the lower and upper geometric CUSUM for detecting a change in p and provided a table of parameters to obtain a specified ARL_0 value. They compared the performance of the CUSUM chart and EWMA chart, based on the number of consecutive conforming items, with the performance of two-stage control chart. It was observed that for the near zero-nonconformity processes, the CUSUM chart and the EWMA chart are more efficient than the two-stage control

Table 7.12 ANIS profile upper-sided geometric Shewhart, geometric CUSUM, Bernoulli CUSUM, and binomial CUSUM charts for p_0 = 0.0001 and p_1 = 0.0005.

p	Geometric Shewhart $H = 1543$	Geometric CUSUM $k = 5493$ $H = 4662$ $\mu = 0$	Bernoulli CUSUM $n = 1$ $k = 1/5493$ $H = 9738/5493$ $\mu = 0$	Binomial CUSUM $n = 101$ $k = 1/54$ $H = 95/54$ $\mu = 0$	Binomial CUSUM $n = 759$ $k = 4/29$ $H = 47/29$ $\mu = 0$
0.00010	69,934.1	69,959.2	69,732.5	69,732.5	69,561.2
0.00015	32,263.6	29,789.4	32,947	32,997.9	33,004.2
0.00020	18,829.0	16,898.6	20,157.0	20,208.1	20,359.3
0.00025	12,496.4	11,196.7	14,128.2	14,176.3	14,406.8
0.00030	8,994.7	8,158.1	10,743.4	10,789.2	11,065.0
0.00035	6,846.1	6,329.1	8,615.3	8,659.8	8,963.0
0.00040	5,427.5	5,129.9	7,167.9	7,211.5	7,532.3
0.00045	4,438.4	4,292.3	6,125.4	6,168.4	6,500.4
0.00050	3,719.1	3,678.4	5,340.4	5,383.6	5,723.0
0.00100	1,271.6	1,445.8	2,293.5	2,339.0	2,682.2
0.00500	200.1	203.1	400.4	450.4	843.9
0.10000	10.0	10.0	20.0	101.0	759.0

chart; the ANNC of the former with a zero initial value is only one-half to one-third of the latter. A similar approach, where the $ANOS_0$ values were specified instead of ARL_0, was given by Chang and Gan (2001) for detecting an increase in p for small values of p_0 (10 – 500 ppm).

Chang and Gan (2001) determined the performance of geometric CUSUM chart and compared with existing CUSUM attribute approaches. Table 7.12 reported the average number of inspection before signal (ANIS) profiles of upper-sided geometric Shewhart, geometric CUSUM, Bernoulli CUSUM, and binomial CUSUM charts with respect to various p.

Szarka and Woodall (2011) gave the condition under which the geometric CUSUM chart and the Bernoulli CUSUM chart are equivalent. Cheng and Yu (2013) proposed a CUSUM chart for monitoring over-dispersed counts in semiconductor industry. The proposed control chart was based on the Poisson-gamma compound distribution for failure mechanism in wafer quality. ARL of one-sided CUSUM chart was calculated using a Markov chain approach and compared with ARLs of the EWMA chart. Results revealed that a CUSUM chart realizes significantly better performance than EWMA chart.

7.2.4 COM–Poisson CUSUM Chart

Suppose that the data for all individuals are available with 100% inspection and let $X_1, X_2, \dots$ are i.i.d. COM–Poisson random variables with p.d.f. defined in (2.21). The following hypotheses are required to test.

$H_0 : \mu = \mu_0$ (Process working at given average number of nonconformities)

$H_1 : \mu \neq \mu_0$ (Process is not working at given average number of nonconformities)

Let μ_0 be the in-control value and μ_1 be the shifted value that we are interested in detecting quickly. For detecting an increase in μ, the μ-CUSUM control statistic is

$$G_i = \text{Max}[0, (X_i - k_G + G_{i-1})] \quad \text{for } i = 1, 2, \dots \tag{7.31}$$

The starting value, G_0, for the statistic is often taken as zero but may be a non-zero value if it desired to give the CUSUM a head start (see Lucas & Crosier, 1982). The choice of k_G may be based on the SPRT analysis of the CUSUM, and the work of Moustakides (1986) has shown that the SPRT-based choice of the *reference value* is optimal in a certain sense. The SPRT analysis of COM–Poisson CUSUM leads to the following expression for k_G.

$$k_G = \frac{\ln [z(\mu_1, v_0)/z(\mu_0, v_0)]}{\ln (\mu_1/\mu_0)} \tag{7.32}$$

The μ-CUSUM chart signals when $G_i > h_G$, where h_G is the upper control limit and is determined based on the prespecified in-control performance. The μ-CUSUM chart generalizes the Bernoulli CUSUM chart, the Poisson CUSUM chart, and the geometric CUSUM chart as special cases, which are discussed in previous section. For example, if $v_0 = \infty$, the reference value k_G reduce to $k_G = \dfrac{-\ln [(1-p_1)/(1-p_0)]}{\ln [p_1(1-p_0)/p_0(1-p_1)]}$, which is the *reference value* of Bernoulli CUSUM chart (see Lee et al. (2013)) with $p = \dfrac{\mu}{1 + \mu}$.

Performance Measure

The performance of COM–Poisson CUSUM chart can be determined using ARL. However, average number of observations to signal was used as the performance measure for the CUSUM charts in the work of Saghir and Lin (2014a). In general, a large ANOS value is desired as long as the process remains in control, while a small ANOS value is preferable if the process is out of control. The measure for ANOS can be based on either zero-state or steady-state assumptions. For zero-state ANOS values, we assume that a shift in the parameter occurs at time zero as the monitoring process is initially set-up or when the chart is in these initial conditions. The steady-state ANOS assumes that the process operates in control for some period of

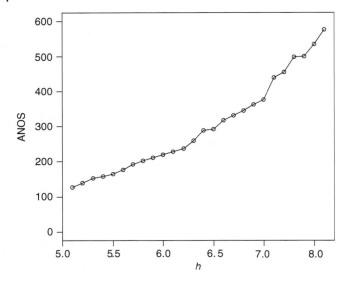

Figure 7.4 Reference value (h) versus ANOS values for the COM–Poisson CUSUM chart.

time and then shifts to an out-of-control state (e.g. see Bourke, 2001). ANOS can be determined via Monte Carlo simulation or Markov chain approach.

Saghir and Lin (2014a) determined the reference value h and zero-state ANOS values using Markov chain approach for various choices of parameter values and in-control false alarm rate. Figure 7.4 gives the reference value h of COM–Poisson CUSUM chart determined by Saghir and Lin (2014a).

7.3 Moving Average (MA) Control Charts for Attributes

An MA control chart is a type of memory control chart based on unweighted MA. Suppose individual observations, $X_1, X_2, \ldots$, are collected, the MA of width w at time i is defined as

$$
M_i = \begin{cases} \dfrac{X_i + X_{i-1} + \cdots + X_{i-w+1}}{w} = \dfrac{\sum\limits_{j=1}^{i-w+1} X_j}{w}, & i \geq w \\[4ex] \dfrac{\sum\limits_{j=1}^{i} X_j}{i}, & i < w \end{cases}
\tag{7.33}
$$

For periods $i < w$, we do not have w observations to calculate an MA of width w. For these periods, the average of all observations up to period i defines the MA. For example, if $w = 3$, then

$$M_1 = \frac{X_1}{1}, M_2 = \frac{X_1 + X_2}{2}, \text{and } M_i = \frac{X_i + X_{i-1} + X_{i-2}}{3}, \quad i \geq 3.$$

If the target value for the process mean is μ_0, then for periods $i \geq w$, the center line and three-sigma control limits are given by

$$\left.\begin{array}{l} \text{LCL} = \mu_0 - 3\dfrac{\sigma}{\sqrt{w}} \\[2mm] \text{CL} = \mu_0 \\[2mm] \text{UCL} = \mu_0 + 3\dfrac{\sigma}{\sqrt{w}} \end{array}\right\} \tag{7.34}$$

whereas for periods $i < w$, $\dfrac{\sigma}{\sqrt{w}}$ is replaced with $\dfrac{\sigma}{\sqrt{i}}$. Here, σ is the in-control standard deviation of the process. Detailed discussions concerning the MA chart are given in Montgomery (2013). When the parameters are unknown, the estimates of mean and variances are used in Eq. (7.34).

7.3.1 Binomial MA Chart

In order to test the proportion nonconforming items at specific level say, $p = p_0$, the binomial MA control chart can also be used as an alternative to Shewhart p-chart. Let the number of nonconforming items in each subgroup denoted by $x_1, x_2, \ldots$. Then, the sample proportion nonconforming of size n for the subgroup i is

$$\hat{p}_i = \frac{x_i}{n} \tag{7.35}$$

The MA of width w at time i is defined as

$$M_i = \begin{cases} \dfrac{\hat{p}_i + \hat{p}_{i-1} + \ldots + \hat{p}_{i-w+1}}{w} = \dfrac{\displaystyle\sum_{j=1}^{i-w+1} \hat{p}_j}{w}, & i \geq w \\[6mm] \dfrac{\displaystyle\sum_{j=1}^{i} \hat{p}_j}{i}, & i < w \end{cases} \tag{7.36}$$

The mean and variance of MA, M_i, are

$$E(M_i) = E\left(\frac{\sum_{j=1}^{i-w+1} \hat{p}_j}{w}\right), \quad i \geq w$$

$$= \frac{1}{w} \sum_{j=i-w+1}^{i} E\left(\hat{p}_j\right)$$

$$= \frac{1}{w}(wp)$$

$$= p$$

and

$$Var(M_i) = Var\left(\frac{\sum_{j=1}^{i-w+1} \hat{p}_j}{w}\right), \quad i \geq w$$

$$= \frac{1}{w^2} \sum_{j=i-w+1}^{i} Var\left(\hat{p}_j\right)$$

$$= \frac{1}{w}\left(\frac{wp(1-p)}{n}\right)$$

$$= \frac{p(1-p)}{wn},$$

respectively. However, for the period $i < w$, the mean and variance are $E(M_i) = p$ and $Var(M_i) = \dfrac{p(1-p)}{in}$, respectively.

When the binomial distribution is adequately approximated by the normal distribution, the three-sigma control chart limits for periods $i \geq w$ can be applied as follows:

$$\left.\begin{aligned} LCL &= p - 3\sqrt{\frac{p(1-p)}{wn}} \\ CL &= p \\ UCL &= p + 3\sqrt{\frac{p(1-p)}{wn}} \end{aligned}\right\} \tag{7.37}$$

For periods $i < w$, the control limits of binomial MA chart are

$$\left.\begin{aligned} LCL &= p - 3\sqrt{\frac{p(1-p)}{in}} \\ CL &= p \\ UCL &= p + 3\sqrt{\frac{p(1-p)}{in}} \end{aligned}\right\} \tag{7.38}$$

Table 7.13 ARL values of binomial MA control chart based on $p = 0.20$ and $n = 100$.

p_1	$w = 2$	$w = 3$	$w = 4$	$w = 5$
0.05	1.02	1.08	1.08	1.08
0.08	1.52	1.53	1.53	1.53
0.10	1.98	1.98	2.02	2.03
0.15	12.64	7.10	5.49	5.05
0.18	117.49	76.88	55.14	42.13
0.20	370.40	370.40	370.40	370.40
0.22	74.00	51.49	38.78	30.91
0.25	8.43	5.72	4.86	4.65
0.28	2.85	2.57	2.65	2.79
0.30	1.98	1.99	2.07	2.13
0.35	1.27	1.29	1.29	1.29
0.40	1.05	1.05	1.05	1.05

Note that if a specified target value p is not available, then replace it with its estimate, $\bar{p}$.

The performance of the binomial MA control chart has been studied by Khoo (2004a) via a mathematical approximation. Table 7.13 gives the ARL values of binomial MA control chart determined by Khoo (2004a).

7.3.2 Poisson MA Chart

Suppose the number of nonconformities in an inspection unit of product, $c_1, c_2, \ldots$, is collected, then the MA of width w at time i can be computed as

$$M_i = \begin{cases} \dfrac{c_i + c_{i-1} + \ldots + c_{i-w+1}}{w} = \dfrac{\sum\limits_{j=1}^{i-w+1} c_j}{w}, & i \geq w \end{cases} \tag{7.39}$$

For periods $i < w$, there are not yet w observations to calculate a Poisson MA of width w. For these periods, the average of all observations up to period i defines the Poisson MA at time i; i.e.

$$M_i = \begin{cases} \dfrac{\sum\limits_{j=1}^{i} c_j}{w}, & i < w \end{cases} \tag{7.40}$$

The mean and variance of PEWMA statistic M_i for $i \geq w$ are

$$E(M_i) = c$$

and

$$\text{Var}(M_i) = \frac{c}{w}, \text{respectively.}$$

The mean and variance of PEWMA statistic M_i for $i < w$ are

$$E(M_i) = c$$

and

$$\text{Var}(M_i) = \frac{c}{i}, \text{respectively.}$$

The three-sigma control limits for the Poisson MA chart for periods $i \geq w$ are

$$\left. \begin{aligned} \text{LCL} &= c - 3\sqrt{\frac{c}{w}} \\ \text{CL} &= c \\ \text{UCL} &= c + 3\sqrt{\frac{c}{w}} \end{aligned} \right\} \tag{7.41}$$

For periods $i < w$, the control limits of binomial MA chart are

$$\left. \begin{aligned} \text{LCL} &= c - 3\sqrt{\frac{c}{i}} \\ \text{CL} &= c \\ \text{UCL} &= c + 3\sqrt{\frac{c}{i}} \end{aligned} \right\} \tag{7.42}$$

Note that if a specified target value c is not available, then replace it with its estimate, $\bar{c} = \dfrac{\sum\limits_{i=1}^{m} c_i}{m}$ from m preliminary samples.

The performance of the Poisson MA control chart has been studied by Khoo (2004b) via simulation study. Table 7.14 gives the ARL values of Poisson MA control chart and c-chart for acceptable mean rate 3 and amount of shift ranges from 0 to 3 determined by Khoo (2004b). Two-sided three-sigma control limits were used in calculations.

The results showed that the Poisson MA chart is a superior alternative to the c-chart, because it provides, in most cases, longer in-control ARLs and shorter out-of-control ARLs. Due to its attractive ARL properties, the Poisson MA control chart should be given high priority by quality control practitioners.

Table 7.14 ARL profiles for Poisson moving average versus *c*-charts for acceptable mean rate of 3.

Poisson mean shift (in standard deviations)	Poisson MA chart			*c*-Chart
	w = 2	*w* = 3	*w* = 4	
0.00	317.2	523.4	455.3	287.7
0.50	46.1	53.0	43.1	63.8
0.75	23.2	25.1	20.3	47.7
1.00	13.4	13.6	11.4	22.0
1.25	8.8	8.6	7.5	13.0
1.50	6.2	6.1	5.4	10.3
2.00	3.7	3.6	3.4	5.6
2.50	2.6	2.5	2.5	3.0
3.00	2.0	2.0	1.9	2.2

7.3.3 Other MA Charts

Khoo and Wong (2008) extended double moving average (DMA) chart with MA of the MA statistic one more time. They proposed this chart with normal observations and also showed the numerical simulations of ARL. According to Khoo and Wong (2008), the performance of the DMA chart is superior to the MA, EWMA, and CUSUM charts for monitoring small and moderate shifts for process mean. Are-epong (2012) developed a simulation approach method for optimal designing of an MA control chart for monitoring the nonconforming product. The characteristic of control chart is ARL, which is the average number of samples taken before an action signal is given. The ARL should be sufficiently large, while the process is still in control and the average delay (AD) time (mean delay of true alarm times) should be small when the process goes out of control. The explicit formulas of ARL and AD for MA-based control chart are presented when observations are from binomial distribution. In particular, the explicit analytical formulas for evaluating ARL and AD are able to get a set of optimal parameters that depend on a width of the MA (*w*) and width of control limit (*H*) for designing MA chart with minimum of AD. Areepong (2013) extended this work for DMA control charts for Poisson distribution. Areepong (2016) proposed an explicit formula for DMA chart of zero-inflated binomial process (DMAZIB). The ARL is a traditional measurement of control chart's performance, the expected number of observations taken from an in-control process until the control chart falsely signals out of control is denoted by ARL_0.

Table 7.15 Optimal design parameters and minimum ARL$_1$ for DMAZIB chart for $p_0 = 0.01$, $\omega = 0.3$.

Shift (δ)	ARL$_0$ = 370			ARL$_0$ = 500		
	k	w	ARL$_1$	k	w	ARL$_1$
0.01	3	21	47.397	3.0905	21	50.284
0.02	3	12	26.811	3.0905	13	28.259
0.03	3	9	18.824	3.0905	10	19.877
0.04	3	7	14.528	3.0905	8	15.260
0.05	3	6	11.824	3.0905	7	12.455
0.06	3	6	10.028	3.0905	6	10.403
0.07	3	5	8.525	3.0905	5	8.908
0.08	3	5	7.624	3.0905	5	7.884
0.09	3	4	6.668	3.0905	4	6.993
0.10	3	4	6.003	3.0905	4	6.237
0.15	3	3	3.985	3.0905	3	4.132
0.20	3	2	1.617	3.0905	3	3.183

Table 7.15 gives an optimal design parameters and minimum ARL$_1$ for DMAZIB chart for $p_0 = 0.01$, $\omega = 0.3$ determined by Areepong (2016).

References

Abujiya, M. R., Farouk, A. U., Lee, M. H., & Mohamad, I. (2013). On the sensitivity of Poisson EWMA control chart. *International Journal of Humanities and Management Sciences*, *1*(1), 18–22.

Areepong, Y. (2012). Optimal parameters of moving average control chart. *International Journal of Applied Physics and Mathematics*, *2*(5), 372–375.

Areepong, Y. (2013). Optimal parameters of double moving average control chart. *International Journal of Mathematical and Computational Sciences*, *7*(8), 1283–1286.

Areepong, Y. (2016). Statistical design of double moving average scheme for zero inflated binomial process. *International Journal of Applied Physics and Mathematics*, *6*(4), 185–193.

Borror, C. M., Champ, C. W., & Rigdon, S. E. (1998). Poisson EWMA control charts. *Journal of Quality Technology*, *30*(4), 352–361.

Bourke, P. D. (1999). Detecting a shift in fraction nonconforming using run-length control charts with 100% inspection. *Journal of Quality Technology*, *23*(3), 225–238.

Bourke, P. D. (2001). Sample size and the binomial CUSUM control chart: The case of 100% inspection. *Metrika, 53*(1), 51–70.

Brook, D., & Evans, D. A. (1972). An approach to the probability distribution of CUSUM run length. *Biometrika, 59*(3), 539–549.

Chang, T. C., & Gan, F. F. (2001). CUSUM charts for high yield processes. *Statistica Sincia, 11*, 791–805.

Cheng, S. S., & Yu, F.-J. (2013). A CUSUM control chart to monitor wafer quality. *International Journal of Industrial and Manufacturing Engineering, 7*(6), 1183–1188.

Dong, Y., Hedayat, A. S., & Sinha, B. K. (2009). Surveillance strategies for detecting change point in incidence rate based on exponentially weighted moving average methods. *Journal of the American Statistical Association, 103*, 843–853.

Gan, F. F. (1990). Monitoring Poisson observations using modified exponentially weighted moving average control charts. *Communications in Statistics-Simulation and Computation, 19*, 103–124.

Gan, F. F. (1993). An optimal design of CUSUM control charts for binomial counts. *Journal of Applied Statistics, 20*(4), 445–460.

Han, S. W., Tsui, K. L., Ariyajunya, B., & Kim, S. B. (2010). A comparison of CUSUM, EWMA, and temporal scan statistics for detection of increases in Poisson rates. *Quality and Reliability Engineering International, 26*, 279–289.

Hawkins, D. M., & Olwell, D. H. (1998). *Cumulative sum charts and charting for quality improvement*. New York: Springer-Verlag.

Joner, M. D., Woodall, W. H., & Reynolds, M. R. (2008). Detecting a rate increase using a Bernoulli scan statistic. *Statistics in Medicine, 27*(14), 2555–2575.

Khoo, M. B. C. (2004a). A moving average control chart for monitoring the fraction non-conforming. *Quality and Reliability Engineering International, 20*, 617–635.

Khoo, M. B. C. (2004b). Poisson moving average versus c chart for nonconformities. *Quality Engineering, 16*(4), 525–534.

Khoo, M. B. C., & Wong, V. H. (2008). A double moving average control chart. *Communications in Statistics – Simulation and Computation, 37*(8), 1696–1708.

Koshti, V. V. (2011). Cumulative sum control charts. *International Journal of Physics and Mathematical Sciences, 1*(1), 28–32.

Kotani, T., Kusukawa, E., & Ohta, H. (2005). Exponentially weighted moving average chart for high-yield processes. *Industrial Engineering and Management Systems, 4*(1), 75–81.

Lee, J., Wang, N., Xu, L., Schuh, A., & Woodall, W. H. (2013). The effect of parameter estimation on upper-sided Bernoulli cumulative sum charts. *Quality and Reliability Engineering International, 29*(5), 639–651.

Lucas, J. M. (1985). Counted data CUSUM's. *Technometrics, 27*, 129–144.

Lucas, J. M., & Crosier, R. B. (1982). Fast initial response for CUSUM quality-control schemes: give your CUSUM a head start. *Technometrics, 24*(3), 199–205.

Lucas, J. M., & Saccucci, M. S. (1990). The exponentially weighted moving average control schemes: Properties and enhancements (with discussion). *Technometrics, 32*, 1–12.

McCool, J. I., & Joyner-Motley, T. (1998). Control charts applicable when the fraction nonconforming is small. *Journal of Quality Technology, 30*(3), 240–247.

Megahed, F. M., Kensler, J. L. K., Bedair, K., & Woodall, W. H. (2011). A note on the ARL of two-sided Bernoulli-based CUSUM control charts. *Journal of Quality Technology, 43*(1), 43–49.

Mei, Y., Han, S. W., & Tsui, K.-L. (2011). Early detecting of a change in Poisson rate after accounting for population size effects. *Statistica Sinica, 21*, 597–624.

Montgomery, D. C. (2013). *Introduction to statistical quality control* (7th ed.). New York, NY: Wiley.

Mousavi, S., & Reynolds, M. R., Jr. (2009). A CUSUM chart for monitoring a proportion with auto correlated binary observations. *Journal of Quality Technology, 41*(4), 401–414.

Moustakides, G. V. (1986). Optimal stopping times for detecting changes in distributions. *Annals of Statistics, 14*, 1379–1387.

Page, E. S. (1962). Cumulative sum schemes using gauging. *Technimetrics, 4*, 97–109.

Perry, M. B., & Pignatiello, J. J. (2011). Estimating the time of step change with Poisson CUSUM and EWMA control charts. *International Journal of Production Research, 49* (10), 2857–2871.

Reynolds, M. R., Jr. (2013). The Bernoulli CUSUM chart for detecting decreases in a proportion. *Quality and Reliability Engineering International, 29*, 529–534.

Reynolds, M. R., Jr., & Stoumbos, Z. G. (2000). A general approach to modeling CUSUM charts for a proportion. *IIE Transactions, 32*(6), 515–535.

Ross, G. J., Adams, N. M., Tasoulis, D. K., & Hand, D. J. (2012). Exponentially weighted moving average charts for detecting concept drift. *Pattern Recognition Letters, 33*, 191–198.

Ryan, A. G., & Woodall, W. H. (2010). Control charts for Poisson count data with baring sample sizes. *Journal of Quality Technology, 42*, 260–274.

Saghir, A., & Lin, A. (2014a). Cumulative sum charts for monitoring the COM-Poisson processes. *Computers and Industrial Engineering, 68*, 65–77.

Saghir, A., & Lin, Z. (2014b). A flexible and generalized EWMA control chart for attributes. *Quality and Reliability Engineering International, 30*(8), 1427–1443.

Sego, L. H., Woodall, W. H., & Reynolds, M. R. (2008). A comparison of surveillance methods for small incidence rates. *Statistics in Mecdicine, 27*(8), 1225–1247.

Shu, L., Jiang, W., & Wu, Z. (2012). Exponentially weighted moving average control charts for monitoring increases in Poisson rate. *IIE Transactions, 44*, 711–723.

Somerville, S. E., Montgomery, D. C., & Runger, G. C. (2002). Filtering and smoothing methods for mixed particle count distributions. *International Journal of Production Research, 40*(13), 2991–3013.

Spliid, H. (2010). An exponentially weighted moving average control chart for Bernoulli data. *Quality and Reliability Engineering International, 26,* 97–113.

Sukparungsee S. (2014). An EWMA *p* chart based on improved square root transformation. *Proceeding of 2014 World Academic of Science, Engineering and Technology,* July 10–11, Prague, Czech Republic.

Sun, J., & Zhang, G. (2000). Control charts based on the number of consecutive conforming items between two successive nonconforming items for the near zero non-conformity processes. *Total Quality Management, 11*(2), 235–250.

Szarka, J. L., III, & Woodall, W. H. (2011). A review and perspective on surveillance of Bernoulli processes. *Quality and Reliability Engineering International, 27,* 735–752.

Testik, M. C., McCullough, B. D., & Borror, C. M. (2006). The effect of estimated parameters on Poisson EWMA control charts. *Quality Technology and Quantitative Management, 3*(4), 513–527.

Weiss, C. H., & Atzmuller, M. (2010). EWMA control charts for monitoring binary processes with applications to medical diagnosis data. *Quality and Reliability Engineering International, 26*(8), 795–805.

White, C. H., Keats, J. B., & Stanley, J. (1997). Poisson CUSUM versus *C* chart for defect data. *Quality Engineering, 9*(4), 673–679.

Woodall, W. H. (2006). The use of control charts in health-care and public-health surveillance. *Journal of Quality Technology, 38*(2), 89–104.

Wu, Z., Jiao, J., & Liu, Y. (2008). A binomial CUSUM chart for detecting large shifts in fraction non-conforming. *Journal of Applied Statistics, 35*(11), 1267–1276.

Yeh, A. B., Mcgrath, R. N., Sembower, M. A., & Shen, Q. (2008). EWMA control charts for monitoring high-yield processes based on non-transformed observations. *International Journal of Production Research, 46*(20), 5679–5699.

Yontay, P., Weiss, C. H., Testik, M. C., & Bayindir, Z. P. (2013). A two-sided cumulative sum chart for first order integer-valued autoregressive processes of Poisson counts. *Quality and Reliability Engineering International, 29,* 33–42.

8

Multivariate Control Charts for Attributes

8.1 Multivariate Shewhart-Type Charts

The multivariate process control is one of the most rapidly developing areas of statistical process control (Woodall and Montgomery 1999). Nowadays, in industry, there are many situations in which the simultaneous monitoring or control of two or more related quality process characteristics is necessary. Monitoring these quality characteristics independently can be very misleading. Process monitoring of problems in which several related variables are of interest is collectively known as *multivariate statistical process control*. The most useful tool of multivariate statistical process control is the quality control chart. A comprehensive review of multi-attribute charts can be found in the studies of Bersimis, Psarakis, and Panaretos (2007), Mason and Young (2002), and Taleb (2009).

8.1.1 Multivariate Binomial Chart

Lu, Xie, Goh, and Lai (1998) proposed a multivariate np- (MNP) chart for monitoring multi-attributes that follow multivariate binomial distribution. For the process being monitored, assume that there are m quality characteristics. Let variable X_i be the number of defects or nonconformities with respect to quality characteristic i, $i = 1, 2, ..., m$. Assume that data $X = (X_1, X_2, ..., X_m)$ follows a jointly m-attribute binomial distribution with p_i – the probability that an item of size n is nonconforming with respect to quality characteristic i. The quality characteristics may not be independent, and we denote the correlation coefficient between characteristic i and characteristic j by r_{ij} such that

$$\begin{cases} r_{ij} = r_{ji} \\ |r_{ij}| \leq 1 \\ r_{ij} = 1, i = j \end{cases}.$$

Introduction to Statistical Process Control, First Edition. Muhammad Aslam, Aamir Saghir, and Liaquat Ahmad.
© 2021 John Wiley & Sons, Inc. Published 2021 by John Wiley & Sons, Inc.

Using a matrix notation, let $\boldsymbol{P} = (p_1, p_2, ..., p_m)$ be the fraction nonconforming vector and $\boldsymbol{\Sigma} = [r_{ij}]_{m \times m}$ be the correlation coefficient matrix. Denote by $\boldsymbol{C} = (C_1, C_2, ..., C_m)$ the vector of counts of nonconforming units, where C_i is the count of nonconforming units with respect to quality characteristic i in a sample. We define a new statistic X, which is the weighted sum of the nonconforming units of all the characteristics in a sample:

$$X = \sum_{i=1}^{m} C_i/\sqrt{p_i}, \quad i = 1, 2, ..., m. \tag{8.1}$$

The quality characteristics of a multivariate attribute process affect the process differently even if these characteristics change by the same amount. More specifically the smaller the fraction nonconforming p_i, the more the count C_i contributes to the statistic X of the multivariate attribute process. Lu et al. (1998) gave two reasons for the choice of $1/\sqrt{p_i}$ as the weight of quality characteristic i's contribution to test statistic X.

The mean and variance of the statistic X can be calculated as follows:

$$
\begin{aligned}
E(X) &= E\left(\sum_{i=1}^{m} C_i/\sqrt{p_i}\right) = \sum_{i=1}^{m} E(C_i)/\sqrt{p_i} \\
&= \sum_{i=1}^{m} np_i/\sqrt{p_i} = n\sum_{i=1}^{m} \sqrt{p_i}
\end{aligned}
\tag{8.2}
$$

and

$$
\begin{aligned}
\mathrm{Var}(X) &= \mathrm{Var}\left(\sum_{i=1}^{m} C_i/\sqrt{p_i}\right) \\
&= \sum_{i=1}^{m} \mathrm{Var}(C_i)/p_i + 2\sum_{i<j} r_{ij}\sqrt{\mathrm{Var}(C_i)\mathrm{Var}(C_j)}/\sqrt{p_i p_j} \\
&= \sum_{i=1}^{m} np_i(1-p_i)/p_i + 2\sum_{i<j} r_{ij}\sqrt{np_i(1-p_i)np_j(1-p_j)}/\sqrt{p_i p_j} \\
&= n\left(\sum_{i=1}^{m}(1-p_i) + 2\sum_{i<j} r_{ij}\sqrt{(1-p_i)(1-p_j)}\right)
\end{aligned}
\tag{8.3}
$$

The correlation coefficient r_{ij} can be either known or estimated from the preliminary samples by the following equation:

$$
\begin{aligned}
r_{ij} &= \mathrm{Cov}(C_i, C_j)/\sqrt{\mathrm{Var}(C_i)\mathrm{Var}(C_j)} \\
&= E\{(X_i - E(X_i))(X_j - E(X_j))\}/\sqrt{\mathrm{Var}(X_i)\mathrm{Var}(X_j)}
\end{aligned}
\tag{8.4}
$$

Following the general structure of Shewhart control chart, the k-sigma limits of the MNP chart are:

$$
\left.\begin{aligned}
\text{LCL} &= n\sum_{i=1}^{m}\sqrt{p_i} - k\sqrt{n\left(\sum_{i=1}^{m}(1-p_i) + 2\sum_{i<j}r_{ij}\sqrt{(1-p_i)(1-p_j)}\right)} \\
\text{CL} &= n\sum_{i=1}^{m}\sqrt{p_i} \\
\text{UCL} &= n\sum_{i=1}^{m}\sqrt{p_i} + k\sqrt{n\left(\sum_{i=1}^{m}(1-p_i) + 2\sum_{i<j}r_{ij}\sqrt{(1-p_i)(1-p_j)}\right)}
\end{aligned}\right\},
$$

$$(8.5)$$

The control chart multiplier k is usually set at 3 but can be adjusted to set the false alarm rate to a specified value, and it sometimes may be advantageous to construct asymmetric control limits. Using the given parameters and control charting constant, the control limits can be computed, and the MNP chart can then be established using Eq. (8.5).

Choice of Sample Size

As for any other control chart an appropriate sample size n should be chosen for its effective application. If the fraction nonconforming vector $\boldsymbol{P}$ is very small, the sample size n should be sufficiently large so that we have a high probability of finding at least one nonconforming unit of a quality characteristic in the sample; otherwise we might find that the control limits are such that the presence of only one nonconforming unit in the sample would indicate an out-of-control condition. To avoid this undesirable situation, we can choose the sample size n so that the probability of finding at least one nonconforming unit per sample is at least equal to some specified value (such as 0.95). Using the cumulative Poisson table in a way similar to univariate np-chart procedures, the condition for the sample size of the MNP chart is

$$
n\sum_{j=1}^{n}p_j \geq 3m
$$

If the in-control value of the fraction nonconforming vector $\boldsymbol{P}$ is small, we can also choose n large enough so that the MNP control chart will have a positive lower control limit (LCL). This ensures that we will have a mechanism to trigger an investigation of one or more samples that contain an unusually small number of nonconforming units. Under the matrix notation, since we wish to have

$$
\text{LCL} = n\sqrt{P}\hat{\imath} - 3\sqrt{n\,\text{Trace}\left(\sqrt{(I-P)(I-P)'\Sigma}\right)} > 0 \tag{8.6}
$$

this implies that

$$n > 9 \frac{\text{Trace}\left(\sqrt{(I-P)(I-P)'\Sigma}\right)}{(\sqrt{P}\hat{I})^2}, \tag{8.7}$$

where $I = [1, 1, 1, ..., 1]_{n \times 1}$ is the unit vector.

Equation (8.7) is easy to compute if the fraction vector nonconforming P and the correlation matrix Σ of a multivariate attribute process are known or have been estimated.

Patel (1973) employed the multivariate normal approximation to the multivariate binomial distribution in order to develop a multivariate scheme to monitor discrete variables, in a similar way that the T^2 statistic monitor the mean vector. Also, Jolayemi (1999) gave a multivariate attribute control chart that is based on an approximation for the convolution of independent variables and extension of the univariate np-chart. Jones, Woodall, and Conerly (1999) proposed a demerit statistic that is a linear combination of the counts of these different types of defects and determined the control limits based on the exact distribution of linear combinations of independent Poisson random variables. Chiu and Kuo (2010) presented a control chart to monitor a bivariate binomial process. The performance of the proposed chart was evaluated by simulation approach and compared with those using both the MNP chart and skewness reduction approaches.

8.1.2 Multivariate Poisson (MP) Chart

Chiu and Kuo (2007) constructed a chart to monitor MP count data, called the MP chart, based on the sum of defects or nonconformities for each quality characteristic as

$$D = \sum_{i=1}^{m} X_i \tag{8.8}$$

They considered that the correlation between the variables comes from a common factor like in Holgate's model, with the difference that they extended it to more than just two variables.

The mean and variance of the statistic D can be calculated as follows:

$$E(D) = E\left(\sum_{i=1}^{m} X_i\right) = \sum_{i=1}^{m} E(X_i)$$
$$= \sum_{i=1}^{m} \lambda_i \tag{8.9}$$

and

$$\text{Var}(X) = \text{Var}\left(\sum_{i=1}^{m} X_i\right)$$

$$= \sum_{i=1}^{m} \text{Var}(X_i) + 2\sum_{i<j} r_{ij}\sqrt{\text{Var}(X_i)\text{Var}(X_j)} \qquad (8.10)$$

$$= \sum_{i=1}^{m} \lambda_i + 2\sum_{i<j} r_{ij}\sqrt{\lambda_i \lambda_j}$$

where λ_i is the defect rate of ith characteristic and r_{ij} is the correlation coefficient between X_i and X_j quality characteristics. The correlation coefficient r_{ij} can be either known or estimated from the preliminary samples by the following equation:

$$r_{ij} = \text{Cov}(X_i, X_j)/\sqrt{\text{Var}(X_i)\text{Var}(X_j)}$$

$$= E\{(X_i - E(X_i))(X_j - E(X_j))\}/\sqrt{\text{Var}(X_i)\text{Var}(X_j)} \qquad (8.11)$$

Following the general structure of Shewhart control chart, the k-sigma limits of the MP chart are

$$\left. \begin{aligned} \text{LCL} &= \sum_{i=1}^{m} \lambda_i - k\sqrt{\sum_{i=1}^{m} \lambda_i + 2\sum_{i<j} r_{ij}\sqrt{\lambda_i \lambda_j}} \\ \text{CL} &= \sum_{i=1}^{m} \lambda_i \\ \text{UCL} &= \sum_{i=1}^{m} \lambda_i + k\sqrt{\sum_{i=1}^{m} \lambda_i + 2\sum_{i<j} r_{ij}\sqrt{\lambda_i \lambda_j}} \end{aligned} \right\}, \qquad (8.12)$$

The control chart multiplier k is usually set at 3 but can be adjusted to set the false alarm rate to a specified value, and it sometimes may be advantageous to construct asymmetric control limits. Using the given parameters and control charting constant, the control limits can be computed, and the MP chart can then be established using Eq. (8.12).

The probability control limits for α level of error of MP chart can be determined as follows:

$$P(D > \text{UCL}) = 1 - \sum_{d=0}^{\text{UCL}-1} \exp\left\{-\left[\sum_{i=1}^{m} \lambda_i - (m-1)r_0\right]\right\}$$

$$\sum_{j=0}^{d/p} \frac{\left(\sum_{i=1}^{m} \lambda_i - mr_0\right)^{d-pj} r_0^j}{(d-pj)!\,j!} \le \frac{\alpha}{2}$$

$$P(D < \text{UCL}) = 1 - \sum_{d=0}^{\text{LCL}} \exp\left\{ -\left[\sum_{i=1}^{m} \lambda_i - (m-1)r_0 \right] \right\}$$
$$\sum_{j=0}^{d/p} \frac{\left(\sum_{i=1}^{m} \lambda_i - mr_0 \right)^{d-pj} r_0^j}{(d-pj)!\,j!} \leq \frac{\alpha}{2}$$

Chiu and Kuo (2007) obtained the simulated average run length (ARL) of the MP control chart for various cases. Table 8.1 gives the simulated ARLs for $p = 2$, shift (δ) 1 (0.25) 6 and = 0.2, 0.3.

Table 8.1 ARL$_1$ for the MP control chart with $p = 2$.

	$\rho = 0.20$		$\rho = 0.50$	
Shift	λ	**ARL$_1$**	λ	**ARL$_1$**
1.00	0.500	491.85	0.5000	217.47
1.25	0.525	369.21	0.5625	141.53
1.50	0.550	279.92	0.6250	89.66
1.75	0.575	202.63	0.6875	62.02
2.00	0.600	169.25	0.7500	44.56
2.25	0.625	138.22	0.8125	33.41
2.50	0.650	115.87	0.8750	27.70
2.75	0.675	96.27	0.9375	22.46
3.00	0.700	82.52	1.0000	18.09
3.25	0.725	71.94	1.0625	15.85
3.50	0.750	60.95	1.1250	13.91
3.75	0.775	55.06	1.1875	11.70
4.00	0.800	49.94	1.2500	10.22
4.25	0.825	42.34	1.3125	9.07
4.50	0.850	37.31	1.3750	8.00
4.75	0.875	36.00	1.4375	7.26
5.00	0.900	31.78	1.5000	6.54
5.25	0.925	29.15	1.5625	6.13
5.50	0.950	26.35	1.6250	5.45
5.75	0.975	23.76	1.6875	4.96
6.00	1.000	22.14	1.7500	4.69

On the other hand, Ho and Costa (2009) analyzed charts on the difference, $DF = X_1 - X_2$, and on the maximum value of two Poisson variables, $MX = \max (X_1, X_2)$. They also assumed Holgate's model. Cozzucoli (2009) proposed a normalized index that can be used to evaluate the capability of the process in terms of the overall level of defectiveness and a two-sided Shewhart-type multivariate control chart to monitor the overall proportion of nonconforming items and the corresponding defectiveness level. Li and Xi (2010) designed a control chart to monitor over-dispersion multivariate count data with either positive or negative correlations. Laungrungrong, Borror, and Montgomery (2011) developed a multivariate exponentially weighted moving average (EWMA) control chart for Poisson variables (the MPEWMA control chart) assuming Holgate's model as well. The MPEWMA chart was compared with the traditional multivariate exponentially weighted moving average (MEWMA) control chart applied to the Poisson variables.

Example 8.1 In this example, we consider the telecommunication data set taken from Jiang, Au, Tsui, and Xie (2002), which consists of 52 samples given in Table 8.2. Using this data set construct an MP chart and interpret it.

Table 8.2 Data set of telecommunication consists of 52 samples from Jiang et al. (2002).

Sample no.	Variable 1	Variable 2	D	Sample No.	Variable 1	Variable 2	D
1	10	0	10	27	17	3	20
2	12	2	14	28	19	1	20
3	12	0	12	29	24	0	24
4	14	3	17	30	17	5	22
5	10	1	11	31	23	3	26
6	14	1	15	32	26	4	30
7	18	4	22	33	17	1	18
8	15	3	18	34	10	1	11
9	11	1	12	35	6	0	6
10	21	3	24	36	11	0	11
11	11	2	13	37	13	1	14
12	18	3	21	38	19	4	23
13	12	7	19	39	11	4	15
14	22	2	24	40	7	3	10
15	13	1	14	41	16	0	16
16	24	4	28	42	17	1	18

(Continued)

Table 8.2 (Continued)

Sample no.	Variable 1	Variable 2	D	Sample No.	Variable 1	Variable 2	D
17	12	2	14	43	11	1	12
18	14	7	21	44	11	2	13
19	18	1	19	45	17	0	17
20	17	4	21	46	21	5	26
21	19	4	23	47	17	2	19
22	11	3	14	48	14	2	16
23	8	3	11	49	10	1	11
24	20	3	23	50	14	3	17
25	23	3	26	51	17	4	21
26	21	3	24	52	11	1	12

Solution

Using the MP chart method, we find the maximum likelihood estimations (MLEs) for the assumed MP distribution as $\hat{\lambda}_1 = 15.31$ and $\hat{\lambda}_2 = 2.35$. The correlation between two variables is about 0.28. The center line of the MP chart is 17.66, and the associated control limits are 5 and 32, respectively (see Figure 8.1). Accordingly, all the points fall within the three-sigma limits, and we can conclude that the bivariate process is statistically in control.

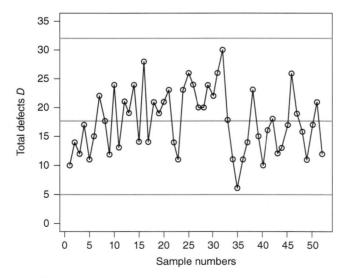

Figure 8.1 The MP chart for the telecommunication data of bivariate counting process.

8.1.3 Multivariate Conway–Maxwell–Poisson (COM–Poisson) Chart

Assume that data $X = (X_1, X_2, ..., X_p)$ follows a jointly m-variate COM–Poisson distribution. Each X_i has a COM–Poisson distribution with the defect rate μ_i that an item is nonconformity and ν_i be dispersion among the number of counts with respect to quality characteristic i. Using a matrix notation, let $\mu = (\mu_1, \mu_2, ..., \mu_p)$ be the average nonconformity vector, $v = (v_1, v_2, ..., v_p)$ be the dispersion vector, and $\rho = [r_{ij}]_{p \times p}$ be the correlation coefficient matrix. Define statistic D as the sum of all X_i:

$$D = \sum_{i=1}^{m} X_i, \quad i = 1, 2, ..., m. \tag{8.13}$$

The mean and variance of the test statistic D can be calculated as follows:

$$
\begin{aligned}
E(D) &= E\left(\sum_{i=1}^{p} X_i\right) = \sum_{i=1}^{p} E(X_i) \\
&= \sum_{i=1}^{p} \mu_i \frac{\partial \log (Z(\mu_i, \, v_i))}{\partial \mu_i}
\end{aligned}
\tag{8.14}
$$

$$\approx \sum_{i=1}^{p} \left[(\mu_i)^{1/v_i} - \frac{v_i - 1}{2v_i} \right] \tag{8.15}$$

and

$$
\begin{aligned}
\mathrm{Var}(D) &= \mathrm{Var}\left(\sum_{i=1}^{p} X_i\right) \\
&= \sum_{i=1}^{p} \mathrm{Var}(X_i) + 2\sum_{i<j} r_{ij} \sqrt{\mathrm{Var}(X_i)\mathrm{Var}(X_j)} \\
&= \sum_{i=1}^{p} \frac{\partial E(X_i)}{\partial \log \mu_i} + 2\sum_{i<j} r_{ij} \sqrt{\frac{\partial E(X_i)}{\partial \log \mu_i} \cdot \frac{\partial E(X_j)}{\partial \log \mu_j}}
\end{aligned}
\tag{8.16}
$$

$$\approx \sum_{i=1}^{p} \frac{1}{v_i} (\mu_i)^{1/v_i} + 2\sum_{i<j} r_{ij} \sqrt{\frac{1}{v_i \cdot v_j} \left[(\mu_i)^{1/v_i} \left(\mu_j\right)^{1/v_j} \right]} \tag{8.17}$$

The correlation coefficient r_{ij} can be either known or estimated from the preliminary samples by the following equation:

$$
\begin{aligned}
r_{ij} &= \mathrm{Cov}(X_i, X_j) / \sqrt{\mathrm{Var}(X_i)\mathrm{Var}(X_j)} \\
&= E\{(X_i - E(X_i))(X_j - E(X_j))\} / \sqrt{\mathrm{Var}(X_i)\mathrm{Var}(X_j)}
\end{aligned}
\tag{8.18}
$$

The control limits of the Shewhart-type MCP control chart, based on exact mean and variance, are

$$\text{UCL} = \sum_{i=1}^{p} \mu_i \frac{\partial \log\left(Z(\mu_i,\ v_i)\right)}{\partial \mu_i} + k\sqrt{\left\{\sum_{i=1}^{p}\frac{\partial E(X_i)}{\partial \log \mu_i} + 2\sum_{i<j}r_{ij}\sqrt{\frac{\partial E(X_i)}{\partial \log \mu_i}\cdot\frac{\partial E(X_j)}{\partial \log \mu_j}}\right\}}$$

$$\text{CL} = \sum_{i=1}^{p} \mu_i \frac{\partial \log\left(Z(\mu_i,\ v_i)\right)}{\partial \mu_i}$$

$$\text{LCL} = \sum_{i=1}^{p} \mu_i \frac{\partial \log\left(Z(\mu_i,\ v_i)\right)}{\partial \mu_i} - k\sqrt{\left\{\sum_{i=1}^{p}\frac{\partial E(X_i)}{\partial \log \mu_i} + 2\sum_{i<j}r_{ij}\sqrt{\frac{\partial E(X_i)}{\partial \log \mu_i}\cdot\frac{\partial E(X_j)}{\partial \log \mu_j}}\right\}}$$

$$(8.19)$$

Note that the approximations provided in Eqs. (8.15) and (8.17) may also be used in the development of control limits in Eq. (8.19) under satisfying conditions. However, the exact terms can be computed easily by using the *COM–Poisson* package in R. Given the values of the parameters, the control limits can be computed, and the MCP chart can then be constructed using Eq. (8.17). If μ_i, v_i, and r_{ij} are unknown, their values can be estimated from the observed k preliminary samples. As a general rule of statistical process control, k should be at least 25. However, the choice of k depends on the performance of a given chart to Phase I analysis.

We plot D on the MCP chart against the sample sequence. To conclude that the process is in control at the present average nonconformity level given as vector $\boldsymbol{\mu} = (\mu_1, \mu_2, ..., \mu_m)$, all the D's should lie inside the control limits, whereas an out-of-control process is signaled by one or more of the D's that exceeds the LCL and upper control limit (UCL). For the proposed MCP chart, the score statistic is defined as

$$Z_i = X_i - E(X_i), \quad i = 1, 2, ..., p. \tag{8.20}$$

If an out-of-control signal is identified, we first compute the score of every individual quality characteristic. Without loss of generality, we define Z_1 and Z_m as the smallest and largest score, respectively. If the statistic D falls above the UCL, Z_m gives the largest positive score, and quality characteristic m is considered as the critical contributor to the upward shift (deterioration) of the process. While the statistic D falls below the LCL, Z_1 gives the smallest negative score, and quality characteristic 1 is considered as the critical contributor to the downward shift (improvement) of the process.

Assume that $X{\sim}CP(\boldsymbol{\mu}, \boldsymbol{v}, \rho)$ is the multivariate COM–Poisson process with in-control average nonconformity and $\boldsymbol{\mu}_s$ is the shifted mean nonconformity vector after shift occurs in $\boldsymbol{\mu}$, then $X{\sim}CP(\boldsymbol{\mu}_s, \boldsymbol{v}, \rho)$. The ARL of multivariate COM–Poisson chart is then defined by:

$$\text{ARL} = \frac{1}{1-\beta},$$

$$\beta = P(\text{LCL} < D < \text{UCL}|\boldsymbol{\mu}_s).$$

where p denotes the probability of statistic D falling within these limits. In case of in-control process, $\boldsymbol{\mu}_s = \boldsymbol{\mu}$. In this section, the ARL will be used to investigate the performance of the proposed MCP chart. Table 8.3 gives the ARL values of the MCP chart determined by Saghir and Lin (2014) for various parameters.

Table 8.3 ARL_1 values of the MCP chart at various dispersion levels $(\mu = 2, \acute{\mu} = 3)$, $p = 2$, and $\text{ARL}_0 \cong 500$.

			$v = 0.85$	$v = 1.0$	$v = 5.0$
r_{12}	δ_1	δ_2	ARL_1	ARL_1	ARL_1
0	1	0	118.5185	102.0408	122.4719
	1	1	39.1010	37.7358	43.1111
	2	0	36.5238	35.9712	39.4712
	2	1	15.0251	14.4928	19.6180
	2	2	7.3333	6.5445	9.0050
0.1	1	0	158.1395	113.6364	166.6667
	1	1	66.2338	39.6825	76.9231
	2	0	58.2857	34.2466	71.4286
	2	1	26.6934	16.0772	33.3333
	2	2	14.0816	7.4627	22.2222
0.2	1	0	142.7805	110.0040	149.2537
	1	1	49.9860	38.4615	55.5556
	2	0	41.3457	37.6667	50.0000
	2	1	21.2735	15.2905	25.0000
	2	2	11.7736	7.2781	15.3846
0.5	1	0	126.3158	115.1144	133.3333
	1	1	51.9048	42.0760	52.6316
	2	0	40.9890	38.1390	43.4783
	2	1	19.5853	17.2674	20.0000
	2	2	11.4843	9.5602	14.2857

Example 8.2   Consider the data set of telecommunication given in Example 8.1. Construct the MCP chart and comments on the process state.

Solution

When using the original bivariate data and applying the optimization scheme to determine the multivariate COM–Poisson MLEs, however we find that $\hat{\lambda}_1 = 9.720$ and $\hat{\nu}_1 = 0.856$ for first quality characteristic and $\hat{\lambda}_2 = 2.160$ and $\hat{\nu}_2 = 0.925$ for second quality characteristic, respectively; i.e. the bivariate data set is over-dispersed as discussed by Li and Xi (2010). Thus, the COM–Poisson is suitable model for fitting the telecommunication bivariate data set because it is flexible to over- or under-dispersed data sets. Therefore, we assume that the data set follows a bivariate COM–Poisson process with mean values of 14.334 and 2.345 and standard deviation values of 4.096 and 1.575, respectively. The correlation between two variables is about 0.28 as considered by Chiu and Kuo (2007). The LCL and UCL of the proposed MCP chart are 2.3313 and 31.0267, respectively, with center line at 16.679. Using these control limits, the MCP chart is constructed in Figure 8.2 for the given data.

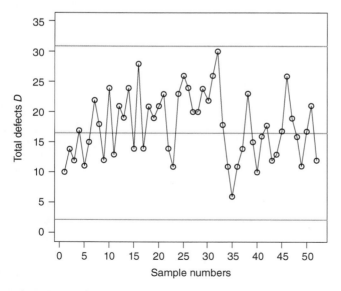

Figure 8.2   The MCP chart for the telecommunication data of bivariate counting process.

The LCL of the proposed MCP chart is wider than the limit of MP chart, while the UCL of MCP chart is contract than the respective limit of the MP chart. From Figure 8.2, all points are within the control limits. Therefore, the bivariate process is statistically in-control status. The same conclusion about the status of the process is obtained with a simple control chart than the two control charts constructed by Jiang's method. At the same time, however, sample 32 would presumably draw attention because it is so close to the upper limit and is beyond an upper warning limit at two standard deviations above the center line (the upper warning bound is 26.244).

Further, this example demonstrates the importance of historical assumptions about distribution. At the same time, if an examiner has reason to believe (as considered by Chiu and Kuo (2007)) that the true model of the data is bivariate Poisson distributed, then the COM–Poisson optimization scheme would produce estimates, $\hat{\lambda}_1 = 15.31$ and $\hat{\nu}_1 = 1$ for first quality characteristic and $\hat{\lambda}_2 = 2.35$ and $\hat{\nu}_2 = 1$ for second quality characteristic, i.e. equivalent to the bivariate Poisson estimates. Accordingly, the underlying assumption produces MCP chart and MP chart control limits that are equivalent. Thus, the results would likewise be equivalent for both chart applications. In this case, however, no such distributional assumption exists because the sample variances of the two variables are 22.884 and 2.897, respectively. It is obviously from an over-dispersed multivariate count process as discussed by Li and Xi (2010). In such case, the application of MP chart is misleading and invalid.

8.2 Multivariate Memory-Type Control Charts

8.2.1 Multivariate EWMA Charts for Binomial Process

An MEWMA chart is one type of multivariate control charts involving a simultaneous monitoring of several correlated quality characteristics. The MEWMA scheme was firstly introduced by Lowry, Woodall, Champ, and Rigdon (1992) as a multivariate version of the univariate EWMA chart for detecting a shift in the mean vectors. In general, the MEWMA scheme is applied to monitor the process changes in the manufacturing industries.

Shamsuzzamana, Haridy, and Zhang (2018) studied a multi-attribute EWMA chart for binomial process. In multi-attribute processes, in which the product has different types of defects, a single multi-attribute chart can be used to monitor the number of nonconforming units found in a sample of size n with respect to the k attributes (or defect types). The overall fraction nonconforming p can be estimated as follows:

$$p = \frac{1}{k} \sum_{j=1}^{k} p_j$$

where p_j represents the fraction nonconforming of the jth attribute. The in-control level of fraction defects is $p_0 = \frac{1}{k} \sum_{j=1}^{k} p_{j0}$, and after δ time change in fraction rate is $p_t = \delta p_0$. The multivariate EWMA chart for monitoring overall fraction defects can be defined as follows:

$$C_t = \lambda(d_t - d_0) + (1 - \lambda)C_{t-1} \tag{8.21}$$

where $C_0 = 0$, λ is EWMA smoothing constant, d_t is the number of nonconformity units found in the tth sample, and d_0 is the in-control value of the product of the sample size n and the in-control fraction nonconforming p_0. An MEWMA chart is implemented as follows:

1) Initialize the statistic C_0 in Eq. (8.21) as zero and set $t = 1$.
2) Take a sample of n units.
3) Determine the number of nonconforming units, d_t, in this sample with respect to all k attributes.
4) Calculate C_t by Eq. (8.21). Determine the critical value H based on the number of attributes, n, and in-control overall level of significance.
5) If $C_t \leq H$, this sample is a conforming one and go back to step (2) to take the next sample. Otherwise (i.e. if $C_t > H$), the current sample is nonconforming and go to step (6).
6) Stop the process immediately for further investigation.

Design of MEWMA Chart

The design of an MEWMA control chart requires the following five specifications:

1) The number k of the attribute characteristics
2) The allowable minimum value τ of ATS_0
3) The individual in-control fraction nonconforming $p_0 j$ of the k attributes (where $j = 1, ..., k$)
4) The maximum shift δ max in overall fraction nonconforming
5) The sample size n

The value of τ is determined with regard to the tolerable false alarm rate. The value of $p_0 j$ for each attribute is usually estimated from the data observed in pilot runs. The maximum shift δ max is required for the calculation of the average number of defectives (AND), which will be discussed shortly. The value of δ max may be chosen based on the knowledge about a process (e.g. the maximum possible p shift in a process) or taken as the shift range the users are interested in.

The sample size n is usually determined according to the available resources such as manpower and measurement instruments.

$$\text{AND} = \frac{N}{\delta_{\max} - 1} \sum_{\delta = 2}^{\delta_{\max}} P_\delta \text{ATS}(P_\delta) \tag{8.22}$$

8.2.2 Multivariate EWMA Charts for Poisson Process

Lowry et al. (1992) presented the MEWMA as an extension to the univariate EWMA. The MEWMA takes into account recent past data that often results in quicker detection of process shifts. In this paper, we examine the MEWMA monitoring technique when it is applied to count data that follows an MP distribution.

Let us say that p quality characteristics, which follow MP distribution, are being monitored simultaneously. The MEWMA statistic is given by

$$\boldsymbol{Z}_t = \boldsymbol{R}\boldsymbol{X}_t + (\boldsymbol{I} - \boldsymbol{R})\boldsymbol{Z}_{t-1} \tag{8.23}$$

where $\boldsymbol{Z}_t$ is the tth MEWMA statistics vector, $\boldsymbol{X}_t$ is the tth observation vector for $t = 1, 2, ..., n$, and $Z_0 = 0$. The vector $\boldsymbol{R}$ consists of weights assigned to past observations in each of the p quality characteristics being monitored and $\boldsymbol{I}$ is the $p \times p$ identity matrix. Specifically, let r_j represent the weight assigned to the jth quality characteristic, then $R = \text{diag}(r_1, r_2, ..., r_2)$, where $0 < r_j \leq 1$ and $j = 1, 2, ..., p$. If equal weight is assigned to each random variable so that $r_1 = r_2 = ... = r_p = \lambda$, then

$$Z_t = RX_t + (1 - \lambda)Z_{t-1}. \tag{8.24}$$

The covariance matrix for the random variable Z_t is $\Sigma_{Z_t} = \left\{ \dfrac{\lambda \left[1 - (1 - \lambda)^{2t} \right]}{2 - \lambda} \right\} \Sigma$,

where Σ is the covariance matrix for the p random variables and is assumed to be known. (Assuming a known covariance matrix is common when evaluating monitoring techniques.) If the covariance matrix is unknown, then it can be estimated using a number of possible methods (see, e.g. Sullivan and Woodall, 1996; Williams, Woodall, Birch, and Sullivan, 2006). As $t \to \infty$, the asymptotic covariance matrix can be written as $\Sigma_{Z_t} = \left\{ \dfrac{\lambda}{2 - \lambda} \right\} \Sigma$.

The MEWMA control chart statistic is given by

$$\boldsymbol{T}_t^2 = \acute{\boldsymbol{Z}}_t \boldsymbol{\Sigma}_{Z_t}^{-1} \boldsymbol{Z}_t. \tag{8.25}$$

An out-of-control signal will occur if $\boldsymbol{T}_t^2 > H$, where $H > 0$ is a threshold limit selected to achieve a desired in-control ARL. The choices of the parameters H and λ can have significant effects on the performance of the MEWMA chart and should

Table 8.4 Control limit (*H*) of MPEWMA chart to achieve ARL$_0$ = 200.

Mean	λ	4 variables		6 variables		10 variables		15 variables	
		$\theta = 1$	$\theta = 0.5$	$\theta = 1$	$\theta = 0.5$	$\theta = 1$	$\theta = 0.5$	$\theta = 1$	$\theta = 0.5$
3	0.10	13.02	13.01	16.68	16.65	23.20	23.17	30.65	30.62
3	0.05	11.49	11.49	14.95	14.93	21.19	21.17	28.41	28.39
5	0.10	12.95	12.95	16.56	16.54	23.03	23.04	30.49	30.49
5	0.05	11.48	11.48	14.92	14.90	21.14	21.15	28.34	28.34
8	0.10	12.692	12.91	16.50	16.50	22.98	22.98	30.41	30.40
8	0.05	11.47	11.47	14.91	14.89	21.13	21.14	28.32	28.31
10	0.10	12.90	12.90	16.50	16.48	22.94	22.95	30.38	30.36
10	0.05	11.46	11.46	14.91	14.89	21.13	21.13	28.31	28.30
15	0.10	12.89	12.89	16.48	16.49	22.94	22.91	30.35	30.33
15	0.05	11.46	11.46	14.90	14.90	21.12	21.12	28.30	28.30

be selected with care. The asymptotic covariance matrix may be used as the covariance matrix of the MPEWMA chart. Table 8.4 gives the control limit *H* to achieve a steady-state ARL of 200 obtained by Laungrungrong et al. (2011).

8.3 Multivariate Cumulative Sum (CUSUM) Schemes

Li and Tsung (2012) proposed the use of multiple binomial and Poisson CUSUM charts with false discovery rate (FDR) control. They proved that the Poisson CUSUM charts with FDR control provide higher detection power and shorter response time. Li and Tsung (2012) provided two methods for obtaining the *p*-values of the Poisson CUSUM statistics, one based on Markov chain with higher precision and the other one based on random walk theory with quicker computation speed. They confirmed that the multiple CUSUM charts with FDR control using either one of the two types of approximation methods lead to better results than the conventional multiple CUSUM charts.

He, He, and Wang (2012) based on the MZIP model proposed by Li et al. (1999) proposed a CUSUM-based procedure for monitoring shifts in a BZIP process where there are only two types of defects considered. Through simulation, they showed that the proposed control charts were effective in detecting shifts in all the parameters of a BZIP process and compared the performance of their proposed method to that of the ZIP CUSUM charts without considering the correlation between the two variables. As a result, He et al. (2012) showed that their method was more sensitive than the ZIP CUSUM charts assuming independence between the two variables.

8.3.1 Multivariate CUSUM Chart for Poisson Data

He, He, and Wang (2014) proposed an MP-CUSUM chart for MP distributed process based on log-likelihood ratio test. Let $Y_1, ..., Y_p$ and Z be the independently Poisson distributed random variables with parameter $\theta_1, \theta_2, ..., \theta_p$ and θ_0, respectively. Then, the data $X = (X_1, X_2, ..., X_p)$ follows a p-variate Poisson distribution that can be constructed as follows (Li et al., 1999):

$$X_1 = Y_1 + Z$$
$$X_2 = Y_2 + Z$$
$$\vdots$$
$$X_p = Y_p + Z$$

We denote X follows a p-dimensional MP distribution as $X \sim \mathbf{MP}(\theta_1, \theta_2, ..., \theta_p, \theta_0)$. The covariance between variables X_i and X_j is

$$\mathbf{Cov}(X_i, X_j) = \theta_0$$

The joint probability density function is given by (Li et al., 1999):

$$P(X) = P(X_1 = x_1, X_2 = x_2, ..., X_p = x_p)$$
$$= \sum_{i=0}^{s} \left(\frac{\theta_0^i e^{-\theta_0}}{i!} \prod_{j=1}^{p} \frac{\theta_j^{x_j - 1} e^{-\theta_j}}{(x_j - 1)!} \right) \tag{8.26}$$

where $s = \min(x_1, x_2, ..., x_p)$.

Marginally, each $X_i (i = 1, 2, ..., p)$ follows Poisson distribution with parameter $\lambda_i = \theta_i + \theta_0$. They develop the MP-CUSUM chart based on this joint distribution.

Let $X_i = (x_{i1}, x_{i2}, ..., x_{ip})$ for $i = 1, 2, ...$ be observations from an MP distributed process. We can calculate the probability of X_i coming from the MP distribution with parameter Θ_0 or Θ_1, respectively, based on Eq. (8.26).

The log-likelihood ratio is defined as

$$R(\mathbf{X}_i | \Theta_0, \Theta_1) = \log \left(\frac{P_1(\mathbf{X}_i)}{P_0(\mathbf{X}_i)} \right)$$
$$= \log \left(\frac{\mathbf{P}(\mathbf{X}_i | \Theta_1)}{\mathbf{P}(\mathbf{X}_i | \Theta_0)} \right) \tag{8.27}$$

The MP-CUSUM statistics S_i for $i = 1, 2, ...$ based on the log-likelihood ratios are defined as

$$S_i = \max(0, S_{i-1} + R(\mathbf{X}_i | \Theta_0, \Theta_1)), \tag{8.28}$$

Table 8.5 ARL values of MP-CUSUM chart with $\rho = 0$.

No.	θ_1	θ_2	θ_0	ρ	δ	$\Theta_1 = 1.2\Theta_0$ $h = 2.75$	$\Theta_1 = 1.5\Theta_0$ $h = 3.556$
1	3	3	1	0.2500	0.0000	200.13	200.18
2	4	3	1	0.2236	0.2667	34.36	44.81
3	3	4	1	0.2236	0.2667	34.36	44.81
4	3	3	2	0.4000	0.4000	15.01	15.82
5	4	4	1	0.2000	0.4000	14.90	16.83
6	3	4	2	0.3651	1.0667	9.27	8.69
7	3	3	2	0.3651	1.0667	9.27	8.69
8	4	4	2	0.3333	1.6000	6.77	5.84
9	5	3	1	0.2041	1.0667	14.26	15.87
10	3	5	1	0.2041	1.0667	14.26	15.87
11	3	3	3	0.5000	1.6000	6.85	5.86
12	5	5	1	0.1667	1.6000	6.69	5.81
13	5	3	3	0.4333	4.2667	4.44	3.49
14	3	5	3	0.4330	4.2667	4.44	3.49
15	5	5	3	0.3750	6.4000	3.39	2.57

where $S_0 = 0$ and the MP-CUSUM chart signal when $S_i > h$ and h is obtained control limit based on the specified in-control ARL. He et al. (2014) presented a Markov chain-based method for approximating the ARL values of the MP-CUSUM chart and a procedure for determining the control limit h as well. Table 8.5 gives some results taken from He et al. (2014). They gave a comparison between the proposed MP-CUSUM chart and the MP chart proposed by Chiu and Kuo (2007) with dimension of $p = 2$ and $p = 3$ and showed that the MP-CUSUM chart had much better performance than the MP chart, especially for small shifts in the parameters.

References

Bersimis, S., Psarakis, S., & Panaretos, J. (2007). Multivariate statistical process control charts: an overview. *Quality and Reliability Engineering International*, 23, 517–543.

Chiu, J. E., & Kuo, T. I. (2007). Attribute control chart for multivariate Poisson distribution. *Communication in Statistics—Theory and Methods*, 37(1), 146–158.

Chiu, J. E., & Kuo, T. I. (2010). Control charts for fraction nonconforming in a bivariate binomial process. *Journal of Applied Statistics*, 37(10), 1717–1728.

Cozzucoli, P. C. (2009). Process monitoring with multivariate *p*-control chart. *International Journal of Quality, Statistics, and Reliability, 2009*, 707583. doi:10.1155/2009/707583

He, S., He, Z., & Wang, G. A. (2012). CUSUM charts for monitoring bivariate zero-inflated Poisson processes with an application in the LED packaging industry. *IEEE Transactions on Components, Packaging and Manufacturing Technology, 2*(1), 169–180.

He, S., He, Z., & Wang, G. A. (2014). CUSUM control charts for multivariate Poisson distribution. *Communications in Statistics—Theory and Methods, 43*(6), 1192–1208.

Ho, L. L., & Costa, A. F. B. (2009). Control charts for individual observations of a bivariate Poisson process. *International Journal of Advanced Manufacturing Technology, 43*, 744–755.

Jiang, W., Au, S. T., Tsui, K. L., & Xie, M. (2002). Process monitoring with univariate and multivariate *c*-charts. Technical Report, The Logistics Institute, Georgia Tech, and The Logistics Institute–Asia Pacific. National University of Singapore.

Jolayemi, J. K. (1999). A statistical model for the design of multivariate control charts. *Indian Journal of Statistics, 61*(2), 351–365.

Jones, L. A., Woodall, W. H., & Conerly, M. D. (1999). Exact properties of demerit control chart. *Journal of Quality Technology, 31*, 207–215.

Laungrungrong, B., Borror, C. M., & Montgomery, D. C. (2011). EWMA control charts for multivariate Poisson-distributed data. *International Journal of Quality Engineering and Technology, 2*(3), 185–211.

Li, C., Lu, J., Park, J., Kim, K., Brinkley, P. A., & Peterson, J. P. (1999). Multivariate zero-inflated Poisson models and their applications. *Technimetrics, 41*(1), 29–38.

Li, Y., & Tsung, F. (2012). Multiple attribute control charts with false discovery rate control. *Quality and Reliability Engineering International, 28*, 857–871.

Li, Y., & Xi, L. (2010). Statistical process control of over-dispersed multivariate count data. IEEExplore.ieee.org/document/5646482.

Lowry, C. A., Woodall, W. H., Champ, C. W., & Rigdon, S. E. (1992). A multivariate exponential weighted moving average control chart. *Technometrics, 34*(1), 46–53.

Lu, X. S., Xie, M., Goh, T. H., & Lai, C. D. (1998). Control chart for multivariate attribute processes. *International Journal of Production Research, 36*, 3347–3489.

Mason, R. L., & Young, J. C. (2002). *Multivariate statistical process control with industrial applications*. The American Statistical Association and the Society for Industrial and Applied Mathematics.

Patel, H. I. (1973). Quality control methods for multivariate binomial and poisson distributions. *Technometrics, 15*(1), 103–112.

Saghir, A., & Lin, Z. (2014). Control chart for monitoring multivariate COM-Poisson attributes. *Journal of Applied Statistics, 41*(1), 200–214.

Shamsuzzamana, M., Haridy, S., & Zhang, L. (2018). Design of multi-attribute EWMA chart. *Proceedings of the International Conference on Industrial Engineering and Operations Management Bandung*, Indonesia (6–8 March 2018).

Sullivan, J. H., & Woodall, W. H. (1996). A comparison of multivariate quality control charts for individual observations. *Journal of Quality Technology, 28*(4), 398–408.

Taleb, H. (2009). Control charts applications for multivariate attribute processes. *Computers and Industrial Engineering, 56*(1), 399–410.

Williams, J. D., Woodall, W. H., Birch, J. B., & Sullivan, J. H. (2006). Distribution of Hotelling's T^2 statistics based on the successive difference estimator. *Journal of Quality Technology, 38*(3), 217–229.

Woodall, W. H., & Montgomery, D. C. (1999). Research issues and ideas in statistical process control. *Journal of Quality Technology, 31*, 376–386.

Appendix A

Areas of the Cumulative Standard Normal Distribution

z	0.00	0.01	0.02	0.03	0.04	z
0.0	0.50000	0.50399	0.50798	0.51197	0.51595	0.0
0.1	0.53983	0.54379	0.54776	0.55172	0.55567	0.1
0.2	0.57926	0.58317	0.58706	0.59095	0.59483	0.2
0.3	0.61791	0.62172	0.62551	0.62930	0.63307	0.3
0.4	0.65542	0.65910	0.62276	0.66640	0.67003	0.4
0.5	0.69146	0.69497	0.69847	0.70194	0.70540	0.5
0.6	0.72575	0.72907	0.73237	0.73565	0.73891	0.6
0.7	0.75803	0.76115	0.76424	0.76730	0.77035	0.7
0.8	0.78814	0.79103	0.79389	0.79673	0.79954	0.8
0.9	0.81594	0.81859	0.82121	0.82381	0.82639	0.9
1.0	0.84134	0.84375	0.84613	0.84849	0.85083	1.0
1.1	0.86433	0.86650	0.86864	0.87076	0.87285	1.1
1.2	0.88493	0.88686	0.88877	0.89065	0.89251	1.2
1.3	0.90320	0.90490	0.90658	0.90824	0.90988	1.3
1.4	0.91924	0.92073	0.92219	0.92364	0.92506	1.4
1.5	0.93319	0.93448	0.93574	0.93699	0.93822	1.5
1.6	0.94520	0.94630	0.94738	0.94845	0.94950	1.6
1.7	0.95543	0.95637	0.95728	0.95818	0.95907	1.7
1.8	0.96407	0.96485	0.96562	0.96637	0.96711	1.8
1.9	0.97128	0.97193	0.97257	0.97320	0.97381	1.9
2.0	0.97725	0.97778	0.97831	0.97882	0.97932	2.0
2.1	0.98214	0.98257	0.98300	0.98341	0.98382	2.1
2.2	0.98610	0.98645	0.98679	0.98713	0.98745	2.2
2.3	0.98928	0.98956	0.98983	0.99010	0.99036	2.3

(Continued)

Introduction to Statistical Process Control, First Edition. Muhammad Aslam, Aamir Saghir, and Liaquat Ahmad.
© 2021 John Wiley & Sons, Inc. Published 2021 by John Wiley & Sons, Inc.

z	0.00	0.01	0.02	0.03	0.04	z
2.4	0.99180	0.99202	0.99224	0.99245	0.99266	2.4
2.5	0.99379	0.99396	0.99413	0.99430	0.99446	2.5
2.6	0.99534	0.99547	0.99560	0.99573	0.99585	2.6
2.7	0.99653	0.99664	0.99674	0.99683	0.99693	2.7
2.8	0.99744	0.99752	0.99760	0.99767	0.99774	2.8
2.9	0.99813	0.99819	0.99825	0.99831	0.99836	2.9
3.0	0.99865	0.99869	0.99874	0.99878	0.99882	3.0
3.1	0.99903	0.99906	0.99910	0.99913	0.99916	3.1
3.2	0.99931	0.99934	0.99936	0.99938	0.99940	3.2
3.3	0.99952	0.99953	0.99955	0.99957	0.99958	3.3
3.4	0.99966	0.99968	0.99969	0.99970	0.99971	3.4
3.5	0.99977	0.99978	0.99978	0.99979	0.99980	3.5
3.6	0.99984	0.99985	0.99985	0.99986	0.99986	3.6
3.7	0.99989	0.99990	0.99990	0.99990	0.99991	3.7
3.8	0.99993	0.99993	0.99993	0.99994	0.99994	3.8
3.9	0.99995	0.99995	0.99996	0.99996	0.99996	3.9

Appendix B

Factors for Constructing Variable Control Charts

n	A	A_2 for $\overline{X}R$	A_2 for median	A_3	c_4	$1/c_4$	B_3	B_4	B_5	B_6	d_2	$1/d_2$
2	2.121	1.881	1.881	2.659	0.798	1.253	0	3.267	0.000	2.606	1.128	0.887
3	1.732	1.023	1.187	1.954	0.886	1.128	0	2.568	0.000	2.276	1.693	0.591
4	1.500	0.729	0.796	1.628	0.921	1.085	0	2.266	0.000	2.088	2.059	0.486
5	1.342	0.577	0.691	1.427	0.940	1.064	0	2.089	0.000	1.964	2.326	0.430
6	1.225	0.483	0.549	1.287	0.952	1.051	0.03	1.97	0.029	1.874	2.534	0.395
7	1.134	0.419	0.509	1.182	0.959	1.042	0.118	1.882	0.113	1.806	2.704	0.370
8	1.061	0.373	0.432	1.099	0.965	1.036	0.185	1.815	0.179	1.751	2.847	0.351
9	1.000	0.337	0.412	1.032	0.969	1.032	0.239	1.761	0.232	1.707	2.970	0.337
10	0.949	0.308	0.362	0.975	0.973	1.028	0.284	1.716	0.276	1.669	3.078	0.325
11	0.905	0.285	0.350	0.927	0.975	1.025	0.321	1.679	0.313	1.637	3.173	0.315
12	0.866	0.266	0.316	0.886	0.978	1.023	0.354	1.646	0.346	1.310	3.258	0.307
13	0.832	0.249	0.307	0.850	0.979	1.021	0.382	1.618	0.374	1.585	3.336	0.300
14	0.802	0.235	0.281	0.817	0.981	1.019	0.406	1.594	0.399	1.563	3.407	0.294
15	0.775	0.223	0.275	0.789	0.982	1.018	0.428	1.572	0.421	1.544	3.472	0.288
16	0.750	0.212	0.255	0.763	0.983	1.017	0.448	1.552	0.440	1.526	3.532	0.283
17	0.728	0.203	0.251	0.739	0.985	1.016	0.466	1.534	0.458	1.511	3.588	0.279
18	0.707	0.194	0.235	0.718	0.985	1.015	0.482	1.518	0.475	1.496	3.640	0.275
19	0.688	0.187	0.231	0.698	0.986	1.014	0.497	1.503	0.490	1.483	3.689	0.271
20	0.671	0.180	0.218	0.680	0.987	1.013	0.51	1.49	0.504	1.470	3.735	0.268
21	0.655	0.173	0.217	0.663	0.988	1.013	0.523	1.477	0.516	1.459	3.778	0.265
22	0.640	0.167	0.210	0.647	0.988	1.012	0.534	1.466	0.528	1.448	3.819	0.262
23	0.626	0.162	0.203	0.633	0.989	1.011	0.545	1.455	0.839	1.438	3.858	0.259
24	0.612	0.157	0.197	0.619	0.989	1.011	0.555	1.445	0.549	1.429	3.895	0.257
25	0.600	0.153	0.191	0.606	0.990	1.010	0.565	1.435	0.559	1.420	3.931	0.254

Introduction to Statistical Process Control, First Edition. Muhammad Aslam, Aamir Saghir, and Liaquat Ahmad.
© 2021 John Wiley & Sons, Inc. Published 2021 by John Wiley & Sons, Inc.

Index

Introduction to Statistical Process Control, First Edition. Muhammad Aslam, Aamir Saghir, and Liaquat Ahmad.
© 2021 John Wiley & Sons, Inc. Published 2021 by John Wiley & Sons, Inc.

Printed and bound by CPI Group (UK) Ltd, Croydon, CR0 4YY